Cristina Popa

Human respiration in wavelenghts

AF167390

Cristina Popa

Human respiration in wavelenghts

Ethylene and ammonia biomarkers in exhaled breath of patients using photoacoustic spectroscopy method

LAP LAMBERT Academic Publishing

Impressum / Imprint

Bibliografische Information der Deutschen Nationalbibliothek: Die Deutsche Nationalbibliothek verzeichnet diese Publikation in der Deutschen Nationalbibliografie; detaillierte bibliografische Daten sind im Internet über http://dnb.d-nb.de abrufbar.

Alle in diesem Buch genannten Marken und Produktnamen unterliegen warenzeichen-, marken- oder patentrechtlichem Schutz bzw. sind Warenzeichen oder eingetragene Warenzeichen der jeweiligen Inhaber. Die Wiedergabe von Marken, Produktnamen, Gebrauchsnamen, Handelsnamen, Warenbezeichnungen u.s.w. in diesem Werk berechtigt auch ohne besondere Kennzeichnung nicht zu der Annahme, dass solche Namen im Sinne der Warenzeichen- und Markenschutzgesetzgebung als frei zu betrachten wären und daher von jedermann benutzt werden dürften.

Bibliographic information published by the Deutsche Nationalbibliothek: The Deutsche Nationalbibliothek lists this publication in the Deutsche Nationalbibliografie; detailed bibliographic data are available in the Internet at http://dnb.d-nb.de.

Any brand names and product names mentioned in this book are subject to trademark, brand or patent protection and are trademarks or registered trademarks of their respective holders. The use of brand names, product names, common names, trade names, product descriptions etc. even without a particular marking in this works is in no way to be construed to mean that such names may be regarded as unrestricted in respect of trademark and brand protection legislation and could thus be used by anyone.

Coverbild / Cover image: www.ingimage.com

Verlag / Publisher:
LAP LAMBERT Academic Publishing
ist ein Imprint der / is a trademark of
OmniScriptum GmbH & Co. KG
Heinrich-Böcking-Str. 6-8, 66121 Saarbrücken, Deutschland / Germany
Email: info@lap-publishing.com

Herstellung: siehe letzte Seite /
Printed at: see last page
ISBN: 978-3-659-61401-9

Human respiration in wavelenghts

Ethylene and ammonia biomarkers in exhaled breath of patients using photoacoustic spectroscopy method

Edited by Dr. Cristina Popa

National Institute for Laser, Plasma and Radiation Physic, Laser Department, 409 Atomistilor St., PO Box MG-36, 077125 Bucharest, Romania

University Politehnica of Bucharest, Faculty of Applied Sciences, Physics Department, 313 Splaiul Independentei, Bucharest - 060042, ROMANIA

Contents

Preface

Since ancient times, physicians have known that human respiration can provide clues to diagnosis of various diseases, e.g., the sweet, fruity odor of acetone indicates uncontrolled diabetes. Researchers have identified over 1,000 different compounds contained in human respiration. These molecules have both endogenous and exogenous origins and provide information about physiological processes occurring in the body as well as environment-related ingestion or absorption of contaminants.

A gaseous molecule that absorbs electromagnetic radiation is excited to a higher electronic, vibrational, or rotational quantum state.

Laser-based photoacoustic detectors are able to monitor trace gas concentrations at atmospheric conditions with orders of magnitude better sensitivity as compared to conventional scientific instrumentation; in addition, they are able to monitor noninvasively and on-line under dynamic conditions.

Trace-gas sensing is a rapidly developing field, in demand for applications such as process and air-quality measurements, atmospheric monitoring, breath diagnostics, biology and agriculture, chemistry, and security and workplace surveillance. Measuring human biomarkers in exhaled breath, is expected to revolutionize diagnosis and management of many diseases and to lead to a rapid, improved, lower-cost diagnosis, which will in turn ensure expanded life spans and an improved quality of life.

The present book focuses on the capabilities of laser photoacoustic spectroscopy method in high impact and new frontier of modern applications: medical diagnosis based on exhaled breath analysis with special importance on the theory, application of the simultaneous measurements of ethylene and ammonia absorptions and the study of exhaled respiration of subjects with renal failure together with breath test analysis of one prevention domain: smoking. This book consists of 4 chapters plus appendices, the first two chapters are focussed on general introduction to human breath biomarkers and the instrumental aspects of the techniques which makes it useful to readers whose main interest is photoacoustic spectroscopy and particularly breath air analysis. The technique developed in this book ensuring the advantages of health state assessment by monitoring the evolution of gaseous biomarkers in human body, impossible to achieve with current techniques. The book also, represents a potentially interesting to researchers and specialists with the application of biomedical optics domain and particularly life sciences area.

Chapter 1.
Biomarkers in human respiration

1.1. Introduction

Biomarkers are present in all parts of the body including body fluids and tissues. Most of the clinical laboratory examinations are done on body fluids such as blood and urine. Biomarkers can be detected on imaging studies or examination of body tissues. Even exhaled breath contains biomarkers. A wide range of technologies is utilized for detection of biomarkers and a number of assays are already available. Biomarkers occur in all parts of the body. For practical purposes, the focus of investigation is on detection of biomarkers in body fluids as most of such specimens can be obtained by non-invasive or minimally invasive techniques. Fluids include blood, urine, and cerebrospinal fluid. Tissue biomarkers can be examined in biopsy specimens of diseases organs. Molecular imaging enables non-invasive in vivo study of biomarkers of disease in various internal organs including the brain. Since most of body fluid biomarkers are proteins, they are described under the section on proteomics. Beyond these conventional sources, biomarkers can also be found in the exhaled breath (Jain, 2010).

In this chapter, are explained the relationship between ethylene (C_2H_4) and different disease and the relationship between ammonia (NH_4) and different disease in human organism.

1.2. What is a biomarker?

The U.S. National Institutes of Health (NIH) defines biomarkers as "physical, functional, or biochemical indicators of a physiologic al or disease process that have diagnostic and/ or prognostic utility." Biomarkers are used in medicine in a number of different ways. One use is to help lead physicians to a diagnosis (Kwak & Preti, 2011).

The application of laser photoacoustic spectroscopy for fast and precise measurements of breath biomarkers has opened up new promises for monitoring and diagnostics in recent years, especially because breath test is a non-invasive method, safe, rapid and acceptable to patients.

The detection of biomarkers in the human breath for the purpose of diagnosis has a long history. Ancient Greek physicians already knew that the aroma of human breath could provide clues to diagnosis. The perceptive clinician was alert for the sweet, fruity odor of acetone in patients with uncontrolled diabetes; the musty, fishy reek of advanced liver disease; the urine-like smell that accompanies failing kidneys; and the putrid stench of a lung abscess. Modern breath analysis is a noninvasive medical diagnostic method that distinguishes among more than 1000 compounds in exhaled breath. These molecules have both endogenous and exogenous origins and provide information about physiological processes occuring in the body as well as environment-related ingestion or absorption of contaminants (Phillips et al., 2003; Phillips et al., 2004; Poli et al., 2005, Charles et al., 2007, Schubert et al., 2005, O'Hara et al., 2009, Risby, 2008).

The bulk matrix of breath is a mixture of nitrogen, oxygen, CO_2 (exhaled in a fraction equal to about four percent of volume), H_2O, and others gases (Table 1.1).

The residual volatile organic compounds (VOC_S) may be endogenous (produced in the body) or exogenous (assimilated as contaminants from the environment). The risk of exposure to VOC_S is defined as "an event that occurs when there is contact at a boundary between a human and the environment with a contaminant of specific concentration for an interval of time" according to the NAS-National Academy of Sciences (Ca et al., 2006). The way for compounds to enter the body is through: ingestion, inhalation, or dermal contact.

Table 1.1. Composition of inhaled and exhaled air

Component		Inhaled air [%]	Exhaled air [%]
nitrogen	●	78.0	78.0
oxygen	●	21.0	15.0
carbon dioxide	●	0.04	4.0
water vapour	●	0.96	3.0

The chemicals can either be absorbed into the systemic blood supply or they can pass through the body not absorbed and be excreted directly in the feces. In general, nonpersistent chemicals with short biological half-lives in blood are eliminated in the urine, or if they are volatile, are eliminated in the exhaled breath air (Ca et al., 2006, Coggiola et al. 2004).

The endogenous compounds found in human breath, such as inorganic gases (e.g., NO and CO), VOCs (e.g., isoprene, ethane, pentane, acetone), and other nonvolatile substances such as isoprostanes, peroxynitrite, or cytokines, can be determined in expired human breath. Tests for endogenous and exogenous biomarkers can provide valuable information concerning a possible disease state or they can indicate recent exposure to drugs or environmental pollutants(Ca et al., 2006, Coggiola et al. 2004, Miekisch et al. 2004).

The most prominent disease marker in exhaled human breath is nitric oxide (NO). It was found that, the average concentration of NO in the breath of healthy humans is 10 to 50 ppb; it is regarded primarily as a noxious gaseous component of air pollution. Intense basic research and clinical investigation have shown that NO is produced by a variety of human tissues. NO is now known to be a central mediator in biological systems and a biomarker for lung cancer, pulmonary hypertension, upper respiratory infection, inflammatory processes in stomach, cancer of digestive organs, acute sepsis, asthma, bronchieactasis and rhinitis (Miekisch et al. 2004, Murtz et al. 2005, Ryter et al. 2007, Wang et al. 2009, Le Marchand et al. 1999).

Breath ethane (C_2H_6) is a biomarker for oxidative stress and destruction caused by free radicals. C_2H_6 (the average concentration in the breath of healthy humans is 0 to 10 ppb) is also an indicator of vitamin E deficiency, cystic fibrosis, ubiquinol status at peroxidation of lipid, smokers and nonsmokers (Skeldon et al. 2006).

There are a number of disease-specific volatile biomarkers which, in general, arise as a direct result of interaction between various metabolic enzymes and their substrates. When the enzymes are defective, commonly due to genetic mutations or non-benign polymorphisms, the substrates (precursors) are not metabolized normally and consequently the disease-specific VOCs are accumulated or reduced (Murtz et al. 2005, Ryter et al. 2007, Wang et al. 2009).

1.3. Breath ethylene in humans

The relation between C_2H_4, free radicals and different disease can be explained by the oxidative stress. In a normal healthy human body, the generation of pro-oxidants in the form of reactive oxygen species (ROS) and reactive nitrogen species (RNS) are effectively kept in check by the various levels of antioxidant defense (Devasagayam et al. 2004). When it gets exposed to psychiatric disorder, adverse physicochemical, environmental or pathological agents, atmospheric pollutants, ultraviolet rays, radiation, toxic chemicals, overnutrition and advanced glycation end products in diabetes, this delicately maintained balance is shifted in favor of pro-oxidants resulting in oxidative stress. It has been implicated in the etiology of several (>100) of human diseases including mental disorders. All the biological molecules present in our body are at risk of being attacked by free radicals. Such damaged molecules can impair cell functions and even lead to cell

5

death, eventually resulting in diseased states. Membrane lipids present in subcellular organelles are highly susceptible to free radical damage. Lipids, when reacted with free radicals, can undergo the highly damaging chain reaction of lipid peroxidation (LP), leading to both direct and indirect effects. During LP, a large number of toxic byproducts are also formed that can have effects at a site away from the area of generation, behaving as "second messengers". The damage caused by LP is highly detrimental to the functioning of the cell. LP is a free radical mediated process. Initiation of a peroxidative sequence is due to the attack by any species, which can abstract a hydrogen atom from a methylene group (CH_2), leaving behind an unpaired electron on the carbon atom ($^{\bullet}CH$). The resultant carbon radical is stabilized by molecular rearrangement to produce a conjugated diene, which then can react with an oxygen molecule to give a lipid peroxyl radical ($LOO^{\bullet}$). These radicals can further abstract hydrogen atoms from other lipid molecules to form lipid hydroperoxides (LOOH) and at the same time propagate LP further. The process of LP, gives rise to many products of toxicological interest including malondialdehyde, 4-hydroxynonenal and a variety of hydrocarbons including pentane, ethane and ethylene (Knight 2000, Devasagayam et al. 2004, Kennedy et al. 2005).

Ethylene (also known as ethene) is a product of LP of linoleic acid and can assess free radical damage (Fig. 1.1). Photoacoustic measurement of the exhaled hydrocarbons, such as ethylene, provides an ideal technique to monitor LP and oxidative stress (Puiu et al. 2007, Kocielnik et al. 2013). Ethylene from the human breath is a marker of oxidant stress and can be directly attributed to biochemical events surrounding LP.

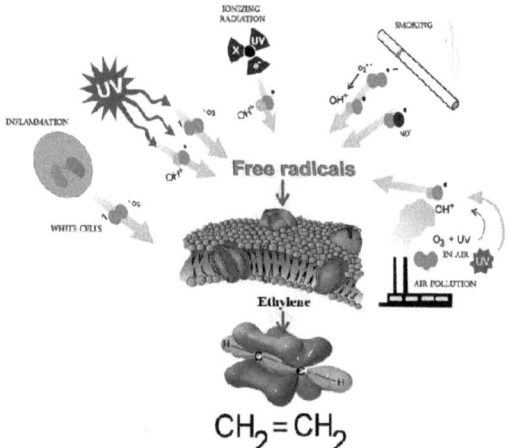

Figure 1.1. Lipid peroxidation and formation of ethylene biomarker.

Also known as ethene, ethylene is a biomarker for acute myocardial infarction, inflammatory processes (chronic asthma, peritonitis), ultraviolet radiation damage of human skin and lipid peroxidation in the lung epithelium. Exhaled concentrations may also be correlated with disease status of non-lung inflammatory diseases such as rheumatoid arthritis and inflammatory bowel disease. Ethylene is vibrationally excited to the $_7$ state, at $= 10.53$ m by the 10P(14) line of the CO_2 laser, where the absorption coefficient is 30.4 cm^{-1}atm^{-1} (Dumitras et al., 2007).

6

1.4. Breath ammonia in humans

Ammonia is produced by all tissues during the metabolism of a variety of compounds, and it is disposed of primarily by formation of urea in the liver (Harvey and Ferrier 2011, Weiner et al. 2011). Elevated blood (breath) ammonia causes pathophysiologic changes (hyperammonemia) in the central nervous system (CNS).

Hyperammonemia is not a true disease, but it is a sign that specific abnormalities may be present that cause blood ammonia to become elevated. There must, therefore, be a metabolic mechanism by which nitrogen is moved from peripheral tissues to the liver for ultimate disposal as urea, while at the same time maintaining low levels of circulating ammonia (Bakouh et al. 2004, Harvey and Ferrier 2011).

Amino acids are quantitatively the most important source of ammonia, because most Western diets are rich in protein and provide excess amino acids, which travel to the liver. However, substantial amounts of ammonia can be obtained from other sources.

The kidneys generate ammonia from glutamine, ammonia beeing excreted into the urine as NH_4^+, which provides an important mechanism for maintaining the body's acid-base balance through the excretion of protons. Ammonia is also obtained from the hydrolysis of glutamine by intestinal *glutaminase*. The intestinal mucosal cells obtain glutamine either from the blood or from digestion of dietary protein.

Ammonia is formed also from urea by the action of bacterial *urease* in the lumen of the intestine. This ammonia is absorbed from the intestine by way of the portal vein and is almost quantitatively removed by the liver via conversion to urea. The urea cycle comprises five enzymes (Fig.1.2): carbamoylphosphate synthetase I, ornithine *trans*carbamylase, argininosuccinate synthetase, argininosuccinatelyase and arginase. For efficient functioning of the pathway in vivo, however, further proteins are required, such as liver glutaminase, mitochondrial carbonic anhydrase V, N-acetylglutamate synthetase, the mitochondrial ornithine and citrulline antiporters and citrin, the mitochondrial aspartate/glutamate antiporter.

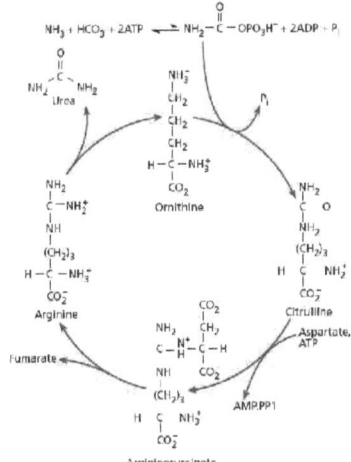

Figure 1.2. The urea cycle (Narasimhan et al.,2001, McCurdy et al., 2007, Dieter, 2011, Brigden et al., 2000).

7

The liver is quantitatively the major organ involved in urea synthesis and it is doubtful whether other cell types, such as enterocytes, can produce significant amounts of urea. However, at least some urea cycle enzymes are found in extrahepatic tissues, where they are involved in providing arginine for nitric oxide synthesis. The initial reaction of the urea cycle is the formation of carbamoyl phosphate from ammonia and bicarbonate, a reaction catalysed by carbamoylphosphate synthetase I, which requires N-acetylglutamate as an allosteric cofactor. Condensation of carbamoyl phosphate with ornithine yields citrulline (by ornithine *trans*carbamylase); this in turn condenses with aspartate to give argininosuccinate (by argininosuccinate synthetase), a reaction that requires the cleavage of two further high-energy phosphate bonds. Argininosuccinate is hydrolysed to fumarate and arginine (by argininosuccinase). Arginine is cleaved by arginase to give urea and ornithine. Ornithine *trans*carbamylase, like carbamoylphosphate synthetase I, is also a major mitochondrial protein; the remaining enzymes are in the cytoplasm of hepatocytes. This necessitates the entry of ornithine into mitochondria and the exit of citrulline, which is brought about by the ornithine/citrulline transporters the mitochondrial ornithine and citrulline antiporters. This series of reactions, returning to ornithine, is known as the urea cycle. It takes part not only in the removal of potentially toxic ammonia, but also in the irreversible removal of bicarbonate. Although the arginase reaction is the major fate of arginine in liver, arginine can also be used for nitric oxide synthesis or undergo decarboxylation to form agmatine. Agmatinase converts arginine to putrescine and urea and the urea is then transported through the blood-stream to be excreted into urine by the kidneys. The reversibility of the process requires an equilibrium concentration of ammonia related to the blood urea nitrogen loading of the blood. As small molecules, ammonia and ammonium ions can penetrate the blood-lung barrier, and appear in exhaled breath. In the case of kidney dysfunction, urea is unable to be excreted, causing an excessive build up of ammonia in the blood. People with kidney failure have a marked odor of ammonia ("fishy") on their breath, which can be an indicator of this disease (Narasimhan et al.,2001, McCurdy et al., 2007).

Although ammonia is constantly produced in the tissues, it is present at very low levels in blood. This is due both to the rapid removal of blood ammonia by the liver (severe impairment of metabolic liver function will produce increased blood ammonia), and the fact that many tissues, particularly muscle, release amino acid nitrogen in the form of glutamine or alanine, rather than as free ammonia.

For example, elevated concentrations of ammonia in the blood cause the symptoms of ammonia intoxication and cell damage, which include: tremors, slurring of speech, somnolence, vomiting, Helicobacter pylori infection, cerebral edema, and blurring of vision. At high concentrations, ammonia can cause coma and death (Handlogten et al. 2005, Harvey and Ferrier 2011).

Healthy kidneys clean the blood by removing excess fluid, minerals, and wastes. They also produce hormones that keep the bones strong and the blood healthy. When the kidneys fail, harmful wastes build up in the body, the blood pressure may rise, the body may retain excess fluid and a decrease in the number of red blood cells can occur. When this happens, haemodialysis treatment is needed to replace the function of the failed kidneys (Stephen and James, 1998).

Ammonia serves as a biomarker for lung cancer, asthma, acute and chronic radiation diseases, metabolism of monoamines in lungs, kidney insufficiency (nephritises, hypertonic disease, atherosclerosis of kidney arteries, toxicosis and nephropatia of pregnants, toxic kidney damage), liver insufficiency at jaundices, hepatitises, liver cirrhosis and toxic hepatitis [8, 11].

A colourless gas with a characteristic pungent odour, NH_3 is vibrationally excited to the $_2$ state, usually by means of the saR (5, K) transitions at $= 9.22$ m. These levels can be excited by the 9R(30) line of the CO_2 laser, where the absorption coefficient has a value of 57.12 $cm^{-1}atm^{-1}$ (Dumitras et al. 2010).

1.5. Collection of exhaled breath samples

The exhaled air is a heterogeneous gas. For a healthy individual, the first part of a exhaled breath, roughly 150 mL, consists of "dead-space" air from the upper airways (such as the mouth and trachea), where air does not come into contact with the alveoli of the lungs. The following part of a breath, about 350 mL, is "alveolar" breath, which comes from the lungs, where gaseous exchange between the blood and breath air takes place. Dead space air can be interpreted as essential for the detection, and depends on the type of molecule detected from the breath test. For example, the dead-space is used to quantify the amount of the NO molecules. In the case of an asthmatic patient, if the airways are inflamed, a high-level of NO is released into the airways and into the dead-space air. But for volatile organic compounds (VOCs) exchanged between blood and alveolar air, the dead-space air is a "contaminant" diluting the concentrations of VOCs when breath air is collected. In terms of the origin of the collected breath gases, there are three basic collection approaches: 1. *upper airway collection* for NO test; this means that only dead-space gas is collected (it is only for the NO test); 2. *alveolar collection*; this means that pure alveolar gas is collected (for tests of other inorganic gases and VOCs); 3. *mixed expiratory collection*; this means that total breath air, including dead-space air and alveolar gas is collected (appropriate for tests of special gases and VOCs). Because the mixed expiratory collection method is easy to perform in spontaneously breathing subjects requiring no additional equipment, it has been most frequently used in practical applications. However, concentrations of endogenous substances in alveolar air are two to three times higher than those found in mixed expiratory samples, because there is no dilution by dead-space gas. Collection of breath air can be performed for a single breath or for collection of individual breathes over a certain period of time. If the sample is collected through a single breath, one has to be sure that this single breath is representative.

How is properly collected a breath sample? To collect a clean breath air sample (Dumitras 2011), we used aluminized multi-patient collection bags (750 mL aluminum-coated bags), composed of a disposable mouthpiece, a tee-mouthpiece assembly (it includes a plastic tee and a removable one-way flutter valve), a discard bag and a tube for nasal breathing (Fig. 1.3).

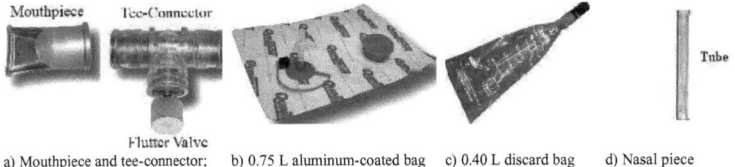

a) Mouthpiece and tee-connector;   b) 0.75 L aluminum-coated bag   c) 0.40 L discard bag   d) Nasal piece

Figure 1.3. The exhaled breath sample collection system (www.QuinTron-USA.com).

Multi-patient collection bags are designed to collect multiple samples from patients and hold a sample for maximum 6 hours.

After an approximately normal inspiration, the subject places the mouthpiece/tube in his mouth/nose, forming a tight seal around it with the lips/nose. A normal expiration is then made through the mouth/nose, in order to empty the lungs of as much air as required to provide the breath sample. For the mouth exhaled breath sample, the first portion of the expired air is directed into the discard bag (with the role in collection of the "dead-space" air: the first portion of an expired breath), while the alveolar air is diverted to the collection bag. When an adequate sample is collected, the subject stops exhaling and removes the mouthpiece/tube.

After the volunteer exhaled via the mouth or the nose and the sample is collected, the gas from the sample is transferred into the PA cell and can be analyzed immediately or later. In either case, it is recommendable to seal the large port with the collection bag port cap furnished with the collection bag. The use of the port cap assures that the sample volume will not be lost due to a leak. Its use also avoids the contamination of the sample by gas diffusion through the one-way valve in the large port, if the sample is stored for a long period of time prior to its analysis.

The correlation between a biomarker and a distinctive disease is often many-fold. In some cases, a breath species is a biomarker that is indicative of about more than one disease or metabolic-disorder; in other cases, one particular disease or metabolic disorder can be characterized by more than one chemical species.

What is the chain of processes involved in a given biomarker? The analysis of exhaled breath to detect or assess a given biomarker with the aim of performing a diagnosis or study a body function is called breath test and can be represented (Fig. 1.4.) in the general form as follows: production of the biomarker during a particular biochemical reaction or a complex metabolic process; diffusion of biomarker through tissues and input into haematic flow; possible intermediate accumulation (buffering); possible trapping of biomarker by utilization and assimilation systems or natural chemical transformation; transport to the lungs; transmembrane diffusion to the air space of lungs; diffusion of biomarker and their mixing with inhaled air in the alveolar space of lungs; release of biomarker in the breathing air; collection of a breath sample and assessment of the biomarker in the breath sample (Popa et al., 2011).

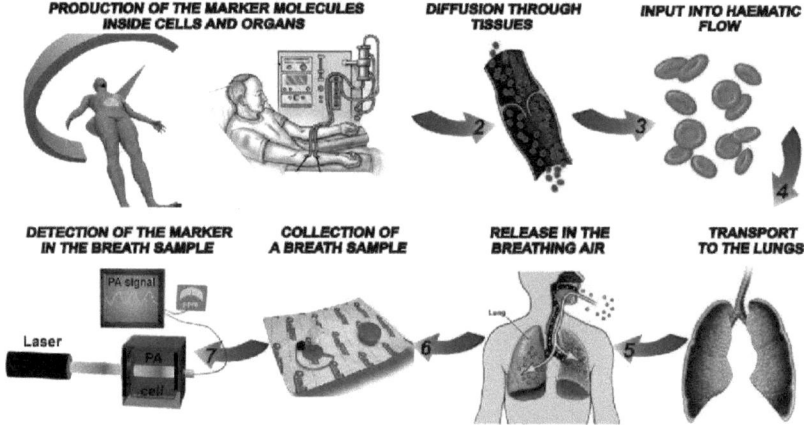

Figure 1.4. The chain of processes involved in a given biomarker.

Breath analysis offers many unique benefits: safe, rapid, simple to perform, non-invasive and frequently repeatable sampling; potential for real-time analyses. Given that human breath contains up to of 200 chemicals, the potential for developing new applications is high. Much of the current knowledge on breath analysis in respiratory medicine derives from years of experience gained in occupational settings, where breath analysis has been used to assess exposure to VOCs.

Chapter 2.
CO₂ laser photoacoustic spectroscopy instrumentation

2.1. Introduction

The photoacoustic/optoacoustic (PA) effect (Harren et al., 2012) was first reported by Alexander Graham Bell (Bell, 1880) when he discovered that thin discs emitted sound when exposed to a rapidly interrupted beam of sunlight. In a later experiment, (Bell, 1881) he removed the eye piece of a commercial spectroscope and placed absorbing substances at the focal point of the instrument. The substances were put in contact with the ear bymeans of a hearing tube and he found 'good' sounds in all parts of the visible and invisible electromagnetic spectrum of the sun. Other publications on this phenomenon followed this first work (Rontgen, 1881, Tyndall, 1881, Preece, 1881).

However, owing to the lack of a quantitative description and the lack of sensitive microphones, the interest in the photoacoustic effect soon declined. In 1938, Viegerov refined the PA technique for the first spectroscopic gas analysis (Viegerov, 1938); there after Pfund (Pfund, 1939) and Luft (Luft, 1943) measured trace gas absorption spectra with an infrared (IR) broadband light source down to the ppm level (part per million). By the end of the 1960s, after the invention of laser, the scientific interest expanded again. In 1968, Kerr and Atwood utilized laser PA detection to obtain the absorption spectrum of small gaseous molecules (Kerr and Atwood, 1968). Owing to the high spectral brightness of lasers and improved phase sensitive lock-in techniques to amplify the acoustic signal, they were able to determine low concentrations of air pollutants. Kreuzer (Kreuzer, 1971) demonstrated that it was possible to detect concentrations of 10 ppbv ($1ppbv = 1:10^9$) of methane in nitrogen, using an intensity-modulated infrared (3 µm) He–Ne laser. Patel demonstrated the potential of the technique by measuring the NO and HO concentrations at an altitude of 28 km with a balloon-borne spin-flip Raman laser.2 (Patel et al., 1974). From hereon, the photoacoustic effect was introduced into the field of trace gas detection with all its environmental, biological, and medical applications.

Laser photoacoustic spectroscopy (LPAS) has emerged over the last decade as a very powerful investigation technique, capable of measuring trace gas concentrations at ppmV (parts per million by volume), or even sub-ppbV (parts per billion by volume) level. LPAS provide several unique advantages, notably the multicomponent capability, high sensitivity and selectivity, wide dynamic range, immunity to electromagnetic interferences, convenient real time data analysis, operational simplicity, relative portability, relatively low cost per unit, easy calibration, and generally no need for sample preparation. CO₂ LPAS offers a sensitive technique for detection and monitoring of trace gases at low concentrations and the spectroscopic system can be adaptable to a broad range of gases and vapors having absorption spectra in the IR with various applications in different disciplines, including nondestructive evaluation of materials, environmental analysis, agricultural, biological, and medical applications, investigation of physical processes (phase transitions, heat and mass transfer, kinetic studies), and many others (Dumitras et al., 2007a,).

In this chapter, are described in detail the components of an instrument based on LPAS principles. Special emphasis is laid on the home-built, frequency-stabilized, line-tunable CO₂-laser source and the resonant photoacoustic cell. Other aspects of a functional photoacoustic instrument, such as the gas handling system and data acquisition and processing, are outlined.

2.2. Block diagram of laser photoacoustic spectrometer

PA spectroscopy is an indirect technique in that an effect of absorption is measured rather than absorption itself. Hence the name of photoacoustic: light absorption is detected through its accompanying acoustic effect. The advantage of PA is that the absorption of light is measured on a zero background; this is in contrast with direct absorption techniques, where a decrease of the source light intensity has to be observed. The spectral dependence of absorption makes it possible to determine the nature of the trace components. The PA method is primarily a calorimetric technique, which measures the precise number of absorbent molecules by simply measuring the amplitude of an acoustic signal. In LPAS the nonradiative relaxation which generates heat is of primary importance. In the IR spectral region, nonradiative relaxation is much faster than radiative decay.
The PA effect in gases can be divided into five main steps (Dumitras et al., 2007a, Dumitras et al., 2012a, Dumitras et al., 2012b):

- Modulation of the laser radiation (either in amplitude or frequency) at a wavelength that overlaps with a spectral feature of the target species; an electrooptical modulation device may also be employed, or the laser beam is modulated directly by modulation of its power supply; the extremely narrowband emission of the laser allows the specific excitation of molecular states; the laser power should be modulated with a frequency in the range $\tau_{th} \gg 1/f \gg \tau_{nr}$, where τ_{th} is the thermal relaxation time, and τ_{nr} the nonradiative lifetime of the excited energy state of the molecule.
- Excitation of a fraction of the ground-state molecular population of the target molecule by absorption of the incident laser radiation that is stored as vibrational-rotational energy; the amount of energy absorbed from the laser beam depends on the absorption coefficient, which is a function of pressure.
- Energy exchange processes between vibrational levels (V-V: vibration to vibration transfer) and from vibrational states to rotational and translational degrees of freedom (V-R, T transfer); the energy which is absorbed by a vibrational-rotational transition is almost completely converted to the kinetic energy of the gas molecules by collisional de-excitation of the excited state; the efficiency of this conversion from deposited to translational energy depends on the pressure and internal energy level structure of the molecule; vibrational relaxation is usually so fast that it does not limit the sensitivity; however, notable anomalies occur in the case of diatomic molecules, such as CO, where vibrational relaxation is slow in the absence of a suitable collision partner, and of the dilute mixtures of CO_2 in N_2, where the vibrational energy is trapped in slowly relaxing vibrational states of N_2; the kinetic energy is then converted into periodic local heating at the modulation frequency.
- Expansion and contraction of the gas in a closed volume that give rise to pressure variation which is an acoustic wave; the input of photon energy with correct timing leads to the formation of a standing acoustic wave in the resonator.
- Monitoring the resulting acoustic waves with a microphone; the efficiency at which sound is transmitted to the microphone depends on the geometry of the cell and the thermodynamic properties of the buffer gas.

From kinetic gas theory it can be estimated that a molecule performs 10^9-10^{10} collisions per second at 1 bar pressure. This means that at atmospheric pressure the photon energy is transformed into an acoustical signal in about 10^{-5}-10^{-6} s. For most polyatomic molecules signal production is even faster. The time needed by the pressure wave to travel from the laser beam area to the microphone in the acoustic cell is therefore in most cases longer than the vibrational relaxation time. For a distance of a few centimeters this transit time is about 10^{-4} s. The time delay between excitation and detection of the pressure wave, however, is influenced not only by energy transfer processes and the transit time, but also by the response time of the gas-microphone system, being about 10^{-4} s or longer (Hess, 1983).

The block diagram of the laser photoacoustic spectrometer for gas studies, is shown in Fig. 2.1.

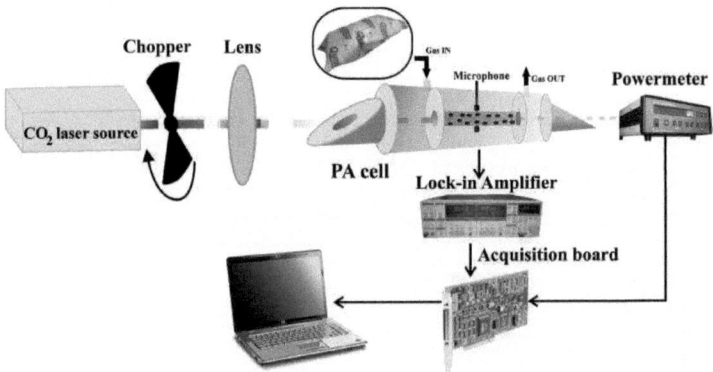

Figure 2.1. Block diagram of laser photoacoustic spectrometer.

The continuous wave, tunable CO_2-laser beam is chopped, focused by a ZnSe lens, and introduced in the PA cell. After passing through the PA cell, the power of the laser beam is measured by a laser radiometer Rk-5700 from Laser Probe Inc. with a measuring head RkT-30. Its digital output is introduced in the data acquisition interface module together with the output from the lock-in amplifier. All experimental data are processed and stored by a computer (Dumitras et al., 2007a).

An advantage of PA spectroscopy as a tool for trace gas analysis is that very few photons are absorbed as the laser beam passes through the sample cell. As a result, notwithstanding the losses from absorption in the windows, the transmitted beam typically has sufficient power for analyzing samples in successive cells, via a multiplexing arrangement. A multiplexed PA sensor can be used to monitor many different samples simultaneously so that one instrument can be deployed to monitor up to 20 different locations within a clean room, industrial plant or other facility (Pushkarsky et al., 2002).

Following the terminology introduced by Miklos et al. (Miklos et al., 2001), the name "PA resonator" will be used for the cavity in which the resonant amplification of the PA signal takes place. The term PA cell (or PA detector; both terms are used in the literature to describe the device in which the PA signal is generated and monitored) is reserved for the entire acoustic unit, including the resonator, acoustic baffles and filters, windows, gas inlets and outlets, and microphone(s). Finally, PA instrument (PA sensor) stands for a complete setup, including the PA cell, light source, gas handling system, and electronics used for signal processing.

It is interesting to mention that the *reverse* PA effect, called "sonoluminiscence", consists in the generation of optical radiation by acoustic waves, while the *inverse* PA effect is the generation of sound due to optical energy being lost from a sample, instead of being deposited in a sample as in the usual PA effect (Tam, 1986).

We use an extracavity arrangement because it has several advantages. In spite of a lower laser power available to excite the absorbing gas in the PA cell, a smaller coherent PA background signal makes it possible to increase the overall sensitivity of the instrument. Also, the dynamic range of the PA method is considerably reduced by intracavity operation. Optical saturation may occur for molecules with high absorption cross section while uncontrollable signal changes may be obtained at higher overall absorption in the PA cell, because the loss of light intensity influences the gain of the laser. This effect may cause erroneous results when the sample concentration changes are large. Therefore, high-sensitivity single-and multipass extracavity PA detectors offer a simpler alternative to intracavity devices.

2.2.1 Mechanical chopper

The light beam was modulated with a high quality, low vibration noise and variable speed (4-4000 Hz) mechanical chopper model DigiRad C-980 or C-995 (30 slot aperture) operated at the appropriate resonant frequency of the cell (564 Hz). The laser beam diameter is typically 5 mm at the point of insertion of the chopper blade and is nearly equal to the width of the chopper aperture. An approximately square waveform was produced with a modulation depth of 100% and a duty cycle of 50% so that the average power measured by the powermeter at the exit of the PA cell is half the cw value. By enclosing the chopper wheel in a housing with a small hole (10 mm) allowing the laser beam to pass, chopper-induced sound vibrations in air that can be transmitted to the microphone detector as noise interference are reduced. A phase reference signal is provided for use with a lock-in amplifier (Dumitras et al., 2007a, Dumitras et al., 2007b, Dumitras et al., 2012a, Dumitras et al., 2012b).

2.2.2 Lock-in amplifier

The generated acoustic waves are detected by microphones mounted in the cell wall, whose signal is fed to a lock-in amplifier locked to the modulation frequency. The lock-in amplifier is a highly flexible signal recovery and analysis instrument, as it is able to measure accurately a single-frequency signal obscured by noise sources many thousands of times larger than itself. It rejects random noise, transients, incoherent discrete frequency interference and harmonics of measurement frequency. A lock-in measures an ac signal and produces a dc output proportional to the ac signal. Because the dc output level is usually greater than the ac input, a lock-in is termed an amplifier. The lock-in can also gauge the phase relationship of two signals at the same frequency. A demodulator, or phase-sensitive detector (PSD), is the basis for a lock-in amplifier. This circuit rectifies the signals coming in at the desired frequency. The PSD output is also a function of the phase angle between the input signal and the amplifier's internal reference signal generated by a phased-locked loop locked to an external reference (chopper). We used a dual-phase, digital lock-in amplifier Stanford Research Systems model SR 830 with the following characteristics: full scale sensitivity, 2 nV - 1 V; input noise, 6 nV (rms)/$\sqrt{Hz}$ at 1 kHz; dynamic reserve, greater than 100 dB; frequency range, 1 mHz – 102 kHz; time constants, 10 μs – 30 s (reference > 200 Hz), or up to 30000 s (reference < 200 Hz).
The diverging IR laser beam is converged by a ZnSe focusing lens (f = 400 mm). In this way, a slightly focused laser beam is passed through the photoacoustic cell without wall interactions (Dumitras et al., 2007a, Dumitras et al., 2012a, Dumitras et al., 2012b).

2.2.3 Line-tunable CO_2 laser

At the Optics and Lasers in Life Sciences, Environment and Manufacturing Laboratory we have designed, constructed and optimized a rugged sealed-off CO_2 laser (named LIR-25 SF), step-tunable on more than 60 vibrational-rotational lines and frequency stabilized by the use of plasma tube impedance variations detected as voltage fluctuations (the optovoltaic method) (Dumitras et al., 1981; Dumitras et al., 1985; Dutu et al., 1985). The glass tube has an inner diameter of 7 mm and a discharge length of 53 cm (Fig. 2.2). At both ends of the tube we attached ZnSe windows at Brewster angle. The laser is water cooled around the discharge tube. The dc discharge is driven by a high-voltage power supply. The end reflectors of the laser cavity are a piezoelectrically driven, partially (85%) reflecting ZnSe mirror at one end and a line-selecting grating (135 lines/mm, blazed at 10.6 µm) at the other. Piezoelectric ceramics such as lead zirconate titanate (PZT) can be used.

Figure 2.2. Homebuilt frequency stabilized CO_2 laser.

The tunability of our CO_2 laser is presented in Fig. 2.3.

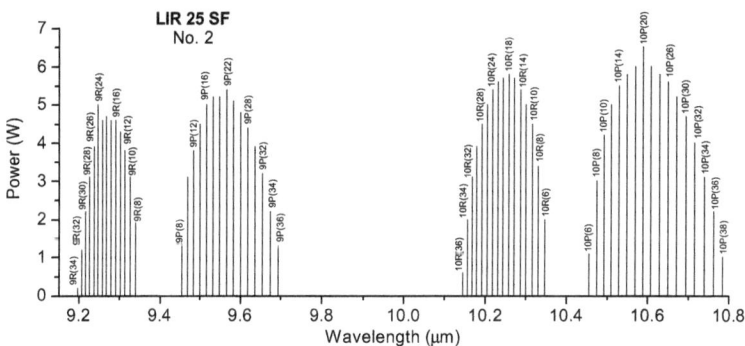

Figure 2.3. Tunability of the low power CO_2 laser with diffraction grating.

We observed the oscillation of 62 different vibrational-rotational lines in both the 10.4 µm and 9.4 µm bands. In this way, the laser was line tunable between 9.2 µm and 10.8 µm with powers varying between 1 and 6.5 W depending on the emitted laser transition. More than 20 lines had output powers in excess of 5 W (Dumitras et al., 2012a, Dumitras et al., 2012b).

2.3. Photoacoustic cell

PA cell is known as the "heart" of photoacoustic spectroscopy and to perform trace gas detection the ideal PA cell should amplify the generated sound origininating from the molecular gas absorption meanwhile rejecting acoustic noise and in-phase IR absorption from other materials.

In the literature the PA cells are often characterized as "nonresonant" or "resonant". This terminology is misleading, because any PA cell can be operated at an acoustic resonance or far from its resonance. Thus, it is preferable to label the system in terms of its nonresonant or resonant mode of operation (Dumitras et al., 2012a, Dumitras et al., 2012b).

To design an optimum acoustically resonant PA cell to be used in CO_2 LPAS, the following requirements have to be met:

- the fraction of laser energy absorbed by the gas must be maximized by increasing either the incident laser power or the optical density of the gas;
- cell responsivity needs to be as high as possible, because the voltage response is proportional to it;
- the microphone responsivity has to be as high as possible, and the use of many microphones is advisable;
- the design must make it possible to operate the cell at an acoustic resonance, and the resonance frequency must lie between 400 and 1000 Hz, where the microphone noise is minimal;
- the quality factor (Q) of the acoustic resonance must not exceed 50 in order to decrease the influence of small deviations from the resonance frequency;
- the electrical noise and the coherent acoustic background noise must be as low as possible; this can be done by using low noise microphones, good acoustic and vibration isolation, low noise electronics, and good electronic isolation;
- the coherent PA background signal due to the heating of the walls and windows must be minimized by using optical components of very high quality and introducing acoustic baffles;
- the cell must enable continuous gas flow operation, and consequently not only the cell windows, but also the gas inlets and outlets have to be positioned at pressure nodes of the resonance;
- the cell must have low gas consumption and fast response, and the cell volume has to be sufficiently small to prevent prohibitive dilution when the produced trace gas is flowed through the cell volume by a continuous gas stream;
- the adsorption and desorption rates on the surfaces in direct contact with the sample gas that can influence particularly measurements on sealed-off samples must be minimized by using special cell materials and reducing the surface-to-volume ratio;
- the effect of the loss mechanisms which we can control must be minimized by an appropriate system design.

Various ways to design (cylindrical geometry, H geometry, T geometry, or using a Helmholtz resonator) and operate resonant PA cells have been studied (Zharov & Letokhov, 1986). Furthermore, PA cells for multipass (Koch & Lahmann, 1978; Nägele & Sigrist, 2000) or intracavity operation (Fung & Lin, 1986; Harren et al., 1990a, Harren et al., 1990b) were designed. The effect of window heating in the amplitude modulation schemes has been minimized by introducing acoustic baffles (Dewey, 1977), developing windowless cells (Gerlach & Amer, 1980; Miklos & Lörincz, 1989; Angeli et al., 1992), or using tunable air columns (Bijnen et al., 1996). In many cases the window-heating signal can be markedly reduced by positioning the entrance and exit of the light beam at nodes of the mode being excited.

A cylindrical cell operated at a radial resonance and having Brewster windows mounted at the pressure nodes of the first radial mode, as presented by Gerlach and Amer (Gerlach & Amer, 1980), does not fulfill all these requirements. Therefore, an open resonant cell excited in its first longitudinal acoustic mode was developed to fulfill most of these requirements.

The H-type longitudinally resonant cell was chosen to form the core of our measuring instrument. Dividing the PA cell into a central chamber and two buffer chambers adjacent to the Brewster windows, a design which lowered significantly the coherent photoacoustic background noise, was first proposed by Tonelli et al. (Tonelli et al., 1983). The characteristics of this type of PA cell have been discussed by Nodov (Nodov, 1978), Kritchman et al. (Kritchman et al., 1978), and Harren et al. (Harren et al., 1990a). Its main advantages are: stable operation at a relatively low frequency; a quality factor of about 20, i.e., much lower than that of a radial resonator, which makes it less sensitive to environmental changes; the efficient conversion of radial to longitudinal modes and the relatively long wavelength guarantee a sufficiently high photoacoustic amplitude; a longitudinal resonator is not noticeably influenced by the gas flow at the desired flow rate of several L/h; noise by gas flow phenomena is negligible for properly positioned inlet and outlet ports; window noise is minimal if the windows are located at a quarter wavelength from the ends of the resonator tube; the construction is rugged and simple and can be achieved with low adhesion materials.

An H-type cylindrical cell designed for resonant PA spectroscopy in gases is shown in Fig.2.4. The longitudinal resonant cell is a cylinder with microphones located at the loop position of the first longitudinal mode (the maximum pressure amplitude). Some general considerations imply that the coherent PA background signal caused by window heating is decreased if the beam enters the cell at the pressure nodes of the resonance. The advantage of mounting the windows at the pressure nodes is well demonstrated, and the window heating signal is decreased by the Q factor. The laser beam enters and exits the cell at the Brewster angle. It is more advantageous to have the beam pass through the windows at the Brewster angle (θ_B), as θ_B is nearly constant over a wide range of wavelengths, and variations of θ_B with wavelength can be tolerated since reflectivity increases very slowly for small deviations from θ_B.

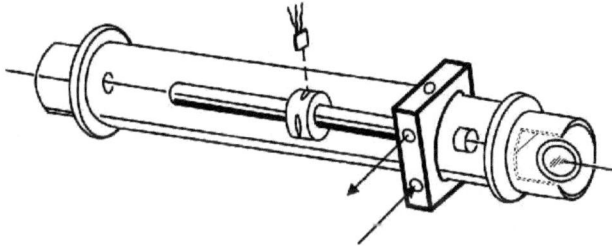

Figure 2.4. Schematic of the PA cell designed for the first longitudinal resonance mode.

The influence of scattered light onto the PA background signal can be minimized by using a highly reflecting polished material, with a good thermally conducting substrate. Bijnen et al. (Bijnen et al., 1996) investigated different materials for the resonant tube and found that the background signal decreased for polished stainless steel, polished brass, and polished, gold-coated copper in a ratio of 6:2:1, respectively. In the case of the CO_2 laser, the best performance was obtained by employing a copper tube with a polished gold coating as resonator material. Because of the excellent heat-conducting properties, the absorbed heat can be quickly dispersed in the copper tube. The gold coating was used not only to optimize laser radiation reflection, but also to obtain a noncorrosive surface to withstand aggressive gases.

Many polar compounds (e.g. ammonia) are highly adsorptive and produce an error in real time concentration measurements by adhering to the detector surfaces. These molecules interact strongly with most metals and many insulating materials. Ammonia is a good model compound for these molecules as it shows the characteristic adsorptive behavior that is not a health hazard at low concentrations. The rate of ammonia adsorption on the gas handling surfaces depends on the surface material and temperature, and on the mixture concentration, flow rate, and pressure. Comparing the ammonia results with those for ethylene, which interacts weakly with most surfaces, provides a measure of the cell-sample interaction. Beck (Beck, 1985) evaluated the suitability of several surface materials for minimizing sample adsorption loss. Four materials–304 stainless steel, gold, paraffin wax, and Teflon–were tested using ammonia as a sample. The results show that both metals interact strongly with the sample. Teflon coating (thickness <25 μm) was found to provide accurate real time response for ammonia sample flows. Also, no signal decay is observed following flow termination. Additionally, the coatings must not degrade the acoustic response of the cell. The Teflon coating actually increases the cell Q by a small amount (1 percent). This is attributed to the smooth slick surface obtained by Teflon coating which would decrease any surface frictional or scattering loss of acoustic energy. Rooth et al. (Rooth et al., 1990) tested the following wall materials – stainless steel 304, gold (on Ni-coated stainless steel), Teflon PTFE, and Teflon PFA – in contact with the gas. Stainless steel proved to be an almost unsaturable reservoir for ammonia at pptV levels. The number of stored molecules exceeded by a factor of 10 or more the number of potential locations on the total geometric surface. Despite its inferior properties in terms of adsorption, Olafsson et al. (Olafsson et al., 1989) used a stainless steel cell and found that an operating temperature of 100°C combined with water vapors led to a very significant reduction of NH_3 adsorption. Apparently, the water molecules stick to the walls even more efficiently than NH_3, and the cell walls are effectively coated with water. Later on, the sample cell was constructed with Teflon as wall material (Olafsson et al., 1992).

Since an open pipe efficiently picks up and amplifies noise from the environment, it should be surrounded by an enclosure. In order to ensure high acoustic reflections at the pipe ends, a sudden change of the cross section is necessary. Therefore, the resonator pipe should open up into a larger volume or to buffers with a much larger cross section. The buffers can be optimized to minimize flow noise and/or window signals. The length of the two buffers accounting for half the resonator length is chosen such as to minimize the acoustic background signal originating from absorption by the ZnSe windows. Open pipes were introduced for PA detection as early as 1977 (Zharov & Letokhov, 1986), and the most sensitive PA detectors currently used are based on open resonant pipes. In resonant cells, window signals can be diminished by using λ/4 buffers next to the windows. These buffers, placed perpendicular to the resonator axis near the windows, are tuned to the resonator frequency and act as interference filters for the window signals (the coupling of the window signals into the resonator is reduced by large buffer volumes that act as interference dampers).

The PA cell is made of stainless steel and Teflon to reduce the outgassing problems and consists of an acoustic resonator (pipe), windows, gas inlets and outlets, and microphones. It also contains an acoustic filter to suppress the flow and window noise. ZnSe windows at the Brewster angle are glued with epoxy (Torr-Seal) to their respective mounts. The resonant conditions are obtained as longitudinal standing waves in an open tube (resonator) are placed coaxially inside a larger chamber. We use an open end tube type of resonator, excited in its first longitudinal mode.

The total cell volume is approximately 1.0 dm^3 (total length 450 mm, and inner diameter 57 mm, minus inner mechanical parts). For flowing conditions, however, it is advantageous to reduce the active volume of the cell. Especially if the flow rate is smaller than 1 L/h (16.6 sccm - standard cubic centimeters per minute), the replenish time for the 1.0 dm^3 cell becomes impractical.

In PA measurements in the gas phase, microphones are usually employed as sensing elements of the acoustic waves generated by the heat deposition of the absorbing molecules. Although high-quality condenser microphones offer the best noise performance, they are rarely used in PA gas detection because of their large size, lower robustness, and relatively high price. The most common microphones employed are miniature electret devices originally developed as hearing aids. The choice of a miniature microphone is particularly advantageous since it can be readily incorporated in the resonant cavity without significantly degrading the Q of the resonance. The frequency response of electret microphones extends beyond 10 kHz, and the response to incident pressure waves is linear over many orders of magnitude.

In our PA cells there are four Knowles electret EK-3033 or EK-23024 miniature microphones in series (sensitivity 20 mV/Pa each at 564 Hz) mounted flush with the wall. They are situated at the loops of the standing wave pattern, at an angle of 90° to one another. The microphones are coupled to the resonator by holes (1 mm diameter) positioned on the central perimeter of the resonator. The battery-powered microphones are mounted in a Teflon ring pulled over the resonator tube. It is of significant importance to prevent gas leakage from inside the resonator tube along the Teflon microphone holder, since minute spacing between the holder and resonator tube produces a dramatic decrease of the microphone signal and the Q value. The electrical output from these microphones is summed and the signal is selectively amplified by a two-phase lock-in amplifier tuned to the chopper frequency.

The resonance curve of our PA cell (cell response in rms volts) was recorded as a function of laser beam chopping frequency and the results are plotted in Fig. 2.5. An accurate method is to construct the resonance curve point by point. In this case, the acoustic signal is measured at different fixed frequencies thus avoiding potential problems arising from the slow formation of a steady state standing wave in the resonator and the finite time resolution of the lock-in amplifier. It is evident from these data that the cell resonance curve is fairly broad, implying that the absorption measurements would not be considerably affected if the frequency of the laser modulation or the cell resonance itself were to shift by a few hertz. PA cells exhibiting narrow resonances require tight control of both temperature and laser modulation frequency to avoid responsivity losses during the experiment.

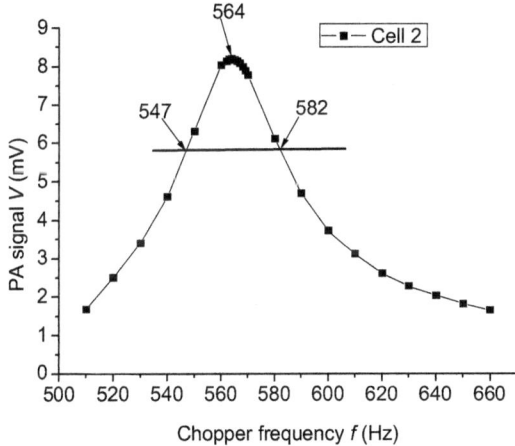

Figure 2.5. Resonance curve for the first longitudinal mode of the PA cell.

The acoustic resonator is characterized by the quality factor Q, which is defined as the ratio of the resonance frequency to the frequency bandwidth between half power points. The amplitude of the microphone signal is $1/\sqrt{2}$ of the maximum amplitude at these points, because the energy of the standing wave is proportional to the square of the induced pressure. The quality factor was

19

measured by filling the PA cell with 1 ppmV of ethylene buffered in nitrogen at a total pressure of 1 atm and by tuning the modulation frequency in 10 Hz increments across the resonance profile to estimate the half width, as described above. For this PA cell, the profile width at half intensity was 35 Hz, yielding a quality factor $Q = 16.1$ at a resonance frequency $f_0 = 564$ Hz. The experimentally determined resonance is not completely symmetric, as the curve rises steeply on one side and becomes less steep on the other side of the maximum. This asymmetry is caused by a coherent superposition of the standing acoustic waves in the detection region of the microphones (Karbach & Hess, 1985).

The calibration of the PA system is usually performed with a reference gas. We calibrated our PA cell with the widely used reference gas ethylene, whose absorption coefficients are accurately known at CO_2-laser wavelengths. Ethylene is well suited for this purpose, since it interacts only weakly with common cell surface materials. Ethylene is chemically inert, has the same molecular weight as nitrogen and possesses no permanent dipole moment which means negligible adsorption on the cell walls. Furthermore, its spectrum within the CO_2-laser wavelength range is highly structured. In particular it exhibits a characteristic absorption peak at the 10P(14) laser transition at 949.49 cm^{-1} which is caused by the proximity of the Q branch of the v_7 vibration of C_2H_4 centered at 948.7715 cm^{-1}. We used a commercially prepared, certified mixture containing 0.96 ppmV C_2H_4 in pure nitrogen throughout our investigations. For calibration we examined this reference mixture at a total pressure p of approximately 1013 mbar and a temperature $T \cong 23°C$ and using the commonly accepted value of the absorption coefficient of 30.4 cm^{-1}atm^{-1} at the 10P(14) line of the $^{12}C^{16}O_2$ laser. For shorter time intervals (by changing the calibration gas mixture after a careful vacuum cleaning of the PA cell), the variation of the cell constant was smaller than 2%. The calibration also depends to some extent on the modulation waveform, since only the fundamental Fourier component of that waveform is resonant with, and hence significantly excites, the first longitudinal mode.

Based on the measured noises, background signals, and cell responsivity, all parameters characterizing the PA instrument can be evaluated (see Table 2.1). Some of them depend on the CO_2 laser and the PA cell, while others are determined by either the coherent acoustic background noise or the coherent PA background signal (Dumitras et al., 2012a, Dumitras et al., 2012b)..

Table 2.1. PA cell parameters.

Parameter/units	Value
Resonance frequency, f_0 (Hz)	564
Quality factor, Q	16.1
Cell responsivity, R (V cm/W)	280
Microphone responsivity, S_M (V/Pa)	$4 \times 20 \times 10^{-3} = 8 \times 10^{-2}$
Cell constant, C (Pa cm/W)	3.5×10^3
Pressure amplitude response, p/P_L (Pa/W)	10^{-1}
Limiting sensitivity of the cell, S_{cell} (W cm^{-1})	2.6×10^{-8}
Limiting sensitivity of the system, S_{sys} (cm^{-1}) (at 4.4 W laser power)	5.9×10^{-9}
Limiting measurable concentration of ethylene, c_{lim} (ppbV)	0.2
Minimum measurable signal in nitrogen, V_{min} (μV) (root mean square)	12
Minimum detectable pressure amplitude, p_{min} (Pa)	4.2×10^{-4}
Minimum detectable concentration, c_{min} (ppbV)	0.89
Minimum detectable absorptivity, α_{min} (cm^{-1})	2.7×10^{-8}
Minimum detectable absorption cross-section per molecule, σ_{min} (cm^2)	1.1×10^{-27}
Cell sensitivity for 1 ppbV of C_2H_4 at 1 W of unchopped laser power, V_{ppb} (μV at 1 ppbV)	3.0

To distinguish the gas absorption signal from other signals (e.g., from the walls, windows, or interfering gases), one has to switch the CO_2 laser to other laser lines. However, repositioning the laser beam to its original wavelength can change the configuration of the laser cavity (deviation in grating position, thermal drift) and result in irreproducible absorption signals if the operation is not carefully conducted. Using a CO_2 laser stabilized on the top of the gain curve ensures that both the laser frequency and output power are reestablished with high accuracy when the laser operation is changed from one line to another (Dumitras et al., 2012a, Dumitras et al., 2012b).

2.4. Gas handling system

The vacuum/gas handling system is an important element in these measurements owing to its role in ensuring PA cell and gas purity. The Teflon/stainless steel system can perform several functions without necessitating any disconnections. It can be used to pump out the cell, mix gases in the desired proportions, and monitor the total pressure of gases. Whenever possible, the PA cell was employed in the gas flow mode of operation to minimize any tendency for the vapor to stick to the cell walls and the effects of the subsequent outgassing of contaminants, which would otherwise lead to increasing background signals during an experimental run (Dumitras et al., 2012a, Dumitras et al., 2012b).

To design an efficient vacuum/gas handling system to be used in LPAS, one must make sure that the following operations can be carried out: evacuation by the vacuum system of the entire gas handling system, including the PA cell, either totally or in different sections; controlled introduction of a gas or gas mixture either for rinsing the PA cell and the gas handling system with pure nitrogen or for calibrating the PA spectrometer with a certified gas mixture; pressure measurement in the PA cell and in different sections of the system; safe insertion in the gas handling system of a sample cuvette (usually made of Pyrex glass) or aluminum-coated plastic bag with the trace gas sample; filtration of certain gases (carbon dioxide and water vapors), which interfere with the trace gas to be measured; controlled introduction of the trace gas to be measured from the sample cuvette or bag into the PA cell by a nonabsorbing gas (nitrogen or synthetic air) acting as carrier; controlled change of the sample and carrier gas flow rates within a broad range (10-1000 sccm); simultaneous measurement of two sample gases (e.g., ethylene and ammonia); quick monitoring of the trace gas concentration in the sample gas by ensuring a response time on the order of minutes or even seconds.

A vacuum/gas handling system to be used in PA experiments was designed and implemented based on these guidelines. The schematic of the gas handling chain is shown in Fig. 2.6.

Gas transport lines throughout the gas mixing station were made of Teflon to minimize adsorption and contamination. The toggle valves V1-V17 and union tees T1-T11 were made of stainless steel. No valve grease was used. The PA cell gas inlet and outlet were connected to the gas handling system with Swagelok fittings. Connections to the inlet and outlet valves of the PA cell were made via flexible Teflon tubing so as to minimize the coupling of mechanical vibrations to the PA cell. The flexible lines also make it possible to position the PA cell during optical alignment.

The pressure of the gases added to the PA cell was determined by means of three Baratron pressure gauges (MKS Instruments, Inc.): 622A (0-1000 mbar), 122A (0-1000 mbar), and 122A (0-100 mbar), connected to a digital two-channel unit PDR-C-2C.

We use thermal mass flowmeters, or mass flow controllers (MFC), to deliver stable and known gas flows to the PA cell. The most critical processes will require flow measurement accuracies of 1% or better in the range 1000 to 10 sccm (7×10^{-4} to 7×10^{-6} mol/sec; 1 sccm (at $0°C$) = 7.436×10^{-7} mol/sec). The digital MFCs sense the mass flow from the temperature difference between two temperature sensors in thermal contact with the gas stream and then process the information digitally with a microcontroller. The analog sensor output is amplified and digitized before it is sent

21

to a microprocessor to compute the final control valve position. The gas flow in our gas handling system is adjusted by two gas flow controllers, MKS 1179A (0 - 1000 sccm) and MKS 2259CC (0 - 200 sccm), which are connected to a digital four-channel instrument MKS 247C.

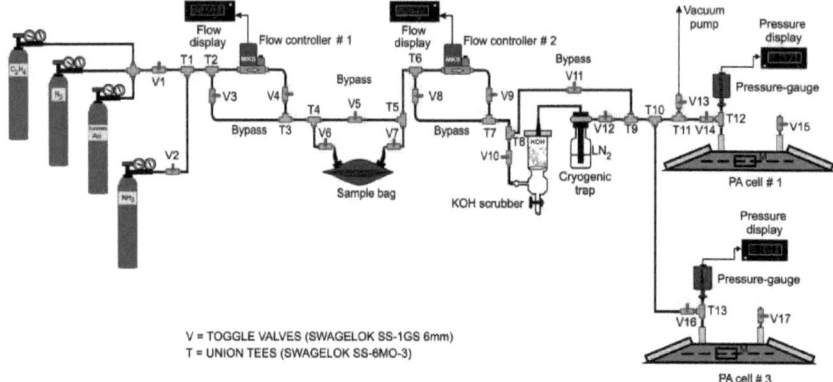

Figure 2.6. Gas handling system.

By using an adequate scrubber for CO_2 filtration, the CO_2 interference problem can be resolved. The CO_2 trap must neither alter the ethylene concentration level, nor introduce new interfering gases.

The following gases were used throughout the experiments:

- ethylene: Linde Gaz Romania, 0.96 ppmV (±5%) C_2H_4 diluted in nitrogen 5.0 (purity 99.999%) and 9.88 ppmV (±2%) C_2H_4 diluted in nitrogen 6.0 (purity 99.9999%)
- nitrogen: Linde Gaz Romania, nitrogen 5.0 (purity 99.999%) and 6.0 (purity 99.9999%);
- synthetic air: Linde Gaz Romania, 20% oxygen and 80% nitrogen (impurities: hydrocarbons max. 0.1 ppmV, nitrogen oxides max. 0.1 ppmV);
- carbon dioxide: Linde Gaz Romania, purity 99.95% (impurities: oils max. 1 ppmV);
- ammonia: Linde Gaz Romania, purity 99.98% and 9.66 ppmV (±5%) NH_3 diluted in nitrogen 6.0 (purity 99.9999%).

The flow rate was usually set at a low value of 30-100 sccm in all experiments in order to eliminate the acoustic noise of the gas flow, and all measurements were carried out with the PA cell at atmospheric pressure. Flow noise increases upwards of 10 L/h (167 sccm) were found to limit the minimum response time of the detector. The flow velocity minimizes the accumulation of the produced gases in the sampling cell. The carrier gas we used was either nitrogen or synthetic air, and its flow rate through the system was monitored by the calibrated flowmeter. Provision is made for bypassing the flowmeter with the gas mixture flow prior to a measurement to equilibrate the feedline surfaces. This ensures that the measured rise times are an exclusive function of the cell characteristics. A measurement is initiated by diverting the gas flow from the bypass through the flowmeter and PA cell and monitoring the PA signal rise that follows (Dumitras et al., 2012a).

As far as the sampling procedure is concerned, we use an extractive method, based on the collection of trace gas samples by some type of container or collecting medium and subsequent analysis in the laboratory. A problem may arise at this point due to some alterations of the gas composition caused by adsorption and desorption processes on the inner surface of the collecting container. The breath samples we analyzed were obtained from volunteers who agreed to provide such samples at certain time intervals. The volunteers were asked to exhale into a sample bag with a normal exhalation flow rate. The breath samples were collected in 0.75-liter aluminum-coated bags (presented in Chapter 1 section 1.5).

2.5. Data acquisition and processing

The acquisition and processing of the recorded data was done with Keithley TestPoint software. TestPoint data acquisition software provides a development environment in which data acquisition applications can be generated. A graphical editor is provided for creating a user interface, or "panel", which the user sees and interacts with as the application executes. A user panel is made of pictorial elements that represent such things as switches, variable controls, numerical, text and selection boxes, bar displays, graphs, and strip charts. In addition, an application editor is provided, which ensures some interactive means of specifying how the visual elements on the user panel interact with the data sources and processing functions to achieve application goals. TestPoint uses an automated textual description of the operations carried out by each user panel element.

We developed a modular software architecture aimed at controlling the experiments, collecting data, and preprocessing information. It helps automate the process of collecting and processing experimental results. The software controls the chopper frequency, transfers powermeter readings, normalizes data, and automatically stores files. It allows the user to set parameters such as the PA cell responsivity, gas absorption coefficient, number of averaged samples at every measurement point, sample acquisition rate, and total number of measurement points. This software interfaces the following instruments: lock-in amplifier; chopper; laser powermeter; gas flowmeter.

The software user interface allows the user to set or read input data and instantaneous values for the PA voltage (rms), average laser power after chopper, and trace gas concentration. Users may set experimental parameters for the PA cell responsivity and gas absorption coefficient. They are also provided with a text input to write a description of the experiments or take other notes. The user interface also provides data visualization.

The software user interface contains three panels which display in real time the following parameters: CO_2 laser power level; PA signal; and trace gas concentration. Another window (countdown) indicates the number of remaining measurement points.

All settings and properties are stored to disk from session to session. In addition, a file may be automatically generated when running an experiment, including: Laser power stores powermeter readings of power incident on the sample as a function of time; PA signal stores the instantaneous values of the PA signal measured by the lock-in amplifier as a function of time; Trace gas concentration stores the time evolution of the trace gas concentration for a given laser wavelength.

A greater clarity on the principles of LPAS is important for understanding the detection of trace gas concentration at sub ppbV level (Dumitras et al., 2007).

Chapter 3.
Measurement of ammonia and ethylene absorption coefficients

3.1. Introduction

In this chapter precise values of absorption coefficients were measured for molecular gases at the wavelengths of the CO_2 laser using a PA cell in an extracavity configuration.

Because more than 250 molecular gases of environmental concern for atmospheric, industrial, medical, military, and scientific spheres exhibit strong absorption bands in the region 9.2 μm – 10.8 μm, we have chosen a frequency tunable CO_2 laser.

Ammonia and ethylene absorption coefficients were measured at both branches of the CO_2 laser lines by using certified gas mixtures and a frequency stabilized laser.

Ethylene and ammonia presents a clear fingerprint spectrum and high absorption strengths in the CO_2 wavelengths region.

3.2. Results

Our laboratory have analyzed the absorption coefficients of ethylene (Dumitras et al., 2007a) and ammonia (Dumitras et al., 2011) at CO_2 laser wavelengths.

The ethylene absorption coefficients for various CO_2 laser transitions have been measured in various experiments. Discrepancies as high as ~15% have been found in the absolute IR values observed at many laser transitions. Such discrepancies are typical of many other gases and are partially associated with the difficulty of producing proper gas samples with known concentration levels. Unfortunately, large discrepancies are also found between measurements of the relative spectral signatures (the ratio between absorption coefficients at different wavelengths). Knowledge of the relative spectral signatures rather than absolute ones is sufficient for trace gas identification. We also note that it is rather problematic to obtain highly accurate measurements of the absolute values of the absorption coefficients of gases using the PA effect. The reason for this is the need of an absolute calibration of the cell. The calibrations cited in the literature are all based on *a priori* knowledge of the absorption coefficient of a gas at some selected wavelength. However, in all cases the absorption coefficient was actually known only to a few percent (Popa et al., 2011a).

Photoacoustics is emerging as a standard technique for measuring extremely low absorptions independent of the path length and offers a degree of parameter control that cannot be attained by other methods. Radiation absorption by the gas creates a pressure signal which is sensed by the microphone. The resulting signal, processed by a phase sensitive detector, is directly proportional to the absorption coefficient and laser power (or laser power absorbed per unit volume). The sensitivity of the technique is such that absorptions of $<10^{-7}$ cm^{-1} can be measured over path lengths of a few tens of centimeters. The small volume of the chamber makes it possible to accurately control the gas parameters, and the system can be operated with static fills or in continuous gas flow mode.

The set of values of the absorption coefficients α, for all laser wavelengths, for a particular gas or vapor and at a common concentration is called the optoacoustic absorption spectrum or signature and is unique to a combination of vapor and laser. These signatures or "fingerprints" are absolute entities, unique only to the laser frequency and species, which provide the specifics of instrument performance in terms of detection limit and interference rejection (Cristescu et al., 2000b).

To improve the measurement of ethylene absorption coefficients, a special procedure was followed. Prior to each run, the gas mixture was flowed at 100 sccm for several minutes to stabilize the boundary layer on the cell walls, since a certain amount of adsorption would occur and possibly influence background signals; after this conditioning period, the cell was closed off and used in measurement. For every gas fill with 0.96 ppmV ethylene buffered in pure nitrogen, the responsivity of the cell was determined supposing an absorption coefficient of 30.4 cm^{-1}atm^{-1} at 10P(14) laser transition. After measurements at all laser lines, the cell responsivity was checked again, to eliminate any possibility of gas desorption during the measurement. The partial pressure of ethylene was enough to have significant PA signals for all laser lines and low enough to be far away from the saturation regime (observations were only made at a C_2H_4 concentration of 100 ppmV). The α values at each laser line were obtained from Eq. (1) using the measured PA signal and laser power (the cell responsivity and ethylene concentration were known).

$$V = \alpha R P_L c. \tag{1}$$

where:
V (V) is the PA signal (peak-to-peak value),
α (cm^{-1} atm^{-1}) is the gas absorption coefficient at a given wavelength,
R (V cm/W) is the (voltage) responsivity of the PA cell or the calibration constant,
P_L (W) is the cw laser power (unchopped value; 2x measured average value) and
c (atm) is the trace gas concentration (usually given in units of per cent, ppmV, ppbV or pptV).
An average over several independent measurements at each line was used to improve the overall accuracy of the results. The values to be presented are thought to be the best published to date.
The absolute magnitudes of the absorption coefficients were calculated as mean values of several independent measurements. An absorption coefficient corresponding to each CO_2 laser transition was determined from two sets of 50 different measurements. Every set of measurements was initiated by the frequency stabilization of a given line of the CO_2 laser. From one set of measurements to another, the closed loop of the frequency stabilization circuit was interrupted, the laser was tuned again to the top of the gain curve, and then the frequency stabilization was set and checked by watching the long term stability. Inside one set, 50 independent measurements were made at a rate of one per second to assess reproducibility. From one measurement to the next, the error measurement of the absorption coefficient was calculated as the ratio between the maximum difference (maximum value minus minimum value) and the average value. The final value of the ethylene absorption coefficient is given by the arithmetic mean of the two sets of measurements, while the absorption coefficient error is chosen as the larger value of the two sets. The same procedure was applied for every absorption coefficient of ethylene/ammonia.
To measure the absorption coefficients of ethylene and ammonia, the software user interface allows recording the laser power, the PA signal and the calculated absorption coefficients on different panels. The multiple measurements in time of the absorption coefficients can be also displayed (Fig. 3.1).

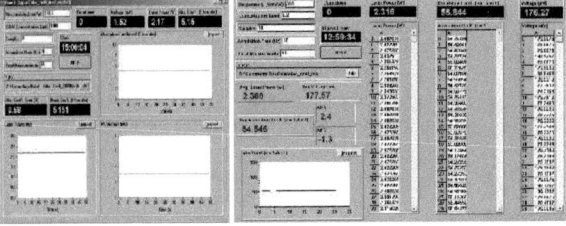

Figure 3. 1. Software user interface to measure the absorption coefficients of ethylene and ammonia.

25

The results of our measurements for ethylene are given in Fig. 3.2. Because of the large spacing between laser transitions (1.2-2 cm^{-1} apart), strong differences of absorption occur. Our results are compared to those of Brewer *et al.* (Brewer et al., 1982) that were also obtained by a photoacoustic method. The difference between the two spectral patterns suggests problems in the measurement techniques (for example, frequency deviation from the laser line center, gas calibration, system purity, linearity, precision) and/or data analysis. The different temperatures and atmospheric pressures at which the measurements were made cannot account for the discrepancies, because Persson *et al.* (Persson et al., 1980) measured a change in absorption coefficient of only 5% at the 10P(14) line for a temperature change of 30°C (negative temperature coefficient), while the changes caused by a pressure difference of 40 Torr are <5% for all CO_2 laser wavelengths.

The random coincidence between the emission and absorption lines will be such that some laser lines will lie close to the centers of the absorbing lines and others will be far away in the wings. The result is a spectral representation unique to that molecule. As a consequence of the superposition of different pressure-broadened C_2H_4 transitions (v_7 vibration), a strong absorption is obtained at the 10P(14) laser line (absorption coefficient of 30.4 $cm^{-1}atm^{-1}$ at 949.479 cm^{-1}). C_2H_4 has weaker absorption coefficients at the 10P(12) and 10P(16) CO_2 laser transitions (4.36 $cm^{-1}atm^{-1}$ at 951.192 cm^{-1} and 5.10 $cm^{-1}atm^{-1}$ at 947.742 cm^{-1}, respectively). Also, in Fig. 3.2 ethylene is seen to possess moderately strong absorption profiles within the 9.4-μm band.

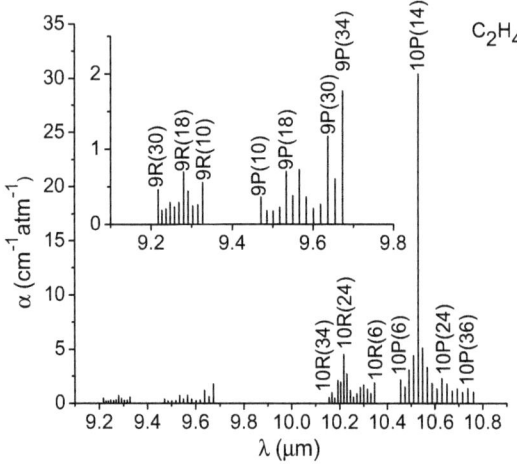

Figure 3.2. Absorption coefficients of ethylene at CO_2 laser wavelengths.

There is general agreement with the results of Brewer et al. (Brewer et al., 1982) for the 00^01-10^00 band. The difference between our results and those obtained by the above mentioned authors is less than 10% for the majority of the investigated lines while only for five lines the discrepancy is higher, between 10% and 20%. By contrast, the values determined in the present work are consistently higher in the 00^01-02^00 band. The difference is larger by 10-50% for the P branch, while our values in the R branch are higher by a factor of 1.5-5.5 (the largest discrepancies are recorded for the 9R(28), 9R(30), and 9R(22) laser lines).

The present work was carried out using a methodology which gave the best possible control over the ethylene partial pressure and background signals. The background levels and calibration of the PA cell were checked before and after every experimental run. The present study is considered reliable, particularly in view of the careful attention that was paid to controlling the gas composition and noise signals. No apparent fault could be found with either the apparatus or methodology that would account for the discrepancy by factors of 2-5 from other reported data in the case of the CO_2 laser 9R lines (Popa et al., 2011a, Popa et al., 2011b, Popa et al., 2011c).

For the measurement of the absorption coefficients of ammonia (Dumitras et al., 2011), the software user interface allows to record the laser power, the PA signal and the calculated absorption coefficients on different panels. The evolution in time of the measurement of the absorption coefficients can also be displayed.

The gas mixture was flowed at 100 sccm for several minutes to stabilize the boundary layer on the cell walls, since a certain amount of adsorption would occur and possibly influence background signals; after this conditioning period, the cell was closed off and used in measurement. For every gas fill with 10 ppm ammonia buffered in pure nitrogen, the responsivity of the cell was determined supposing an absorption coefficient of 57.12 cm^{-1}atm^{-1} at 9R(30) laser transition. This is in accordance both to the measurements reported by Brewer & Bruce (Brewer & Bruce, 1978) and by our tests, when the responsivity of the PA system was checked by measuring the well known absorption coefficient of ethylene at 10P(14) line of the CO_2 laser. After measurements at all laser lines, the cell responsivity was checked again, to eliminate any possibility of gas desorption during the measurement. The α values at each laser line were obtained by using the measured PA signal and laser power and by knowing precisely the ammonia concentration (10.6 ppm) and the responsivity of the PA cell. An average over several independent measurements at each line was used to improve the overall accuracy of the results.

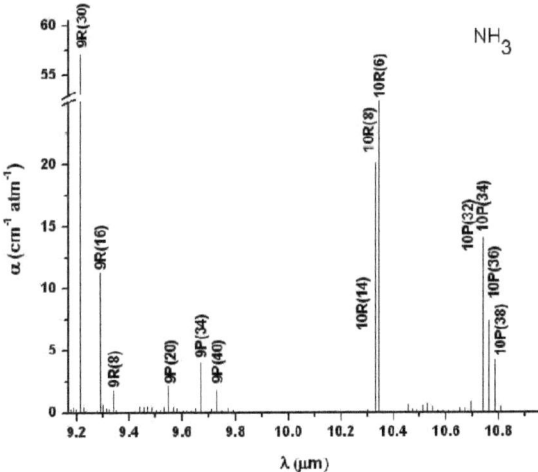

Figure 3.3. Absorption coefficients of ammonia at CO_2 laser wavelengths.

The results of our measurements for ammonia are given in Fig. 3.3. The experimental results show a spectral representation unique to the ammonia molecule. As it can be seen from Fig. 3.3, ammonia has weaker absorption coefficients at other CO_2 laser transitions; some other significant values for the absorption coefficient were found for 9R and 9P bands: 9R(16) - = 11.29 $cm^{-1}atm^{-1}$ (error ± 1.4%), 9P(20) - = 2.10 $cm^{-1}atm^{-1}$ (error ± 2%) and 9P(34) - = 3.99 $cm^{-1}atm^{-1}$ (error ± 0.62%). In the 10R band the measurements gave: 10R(14) - = 6.17 $cm^{-1}atm^{-1}$ (error ± 1.5%), 10R(8) - = 20.08 $cm^{-1}atm^{-1}$ (error ± 1.3%), 10R(6) - = 26.2 $cm^{-1}atm^{-1}$ (error ± 1.7%), and for the 10P band: 10P(32) - = 12.45 $cm^{-1}atm^{-1}$ (error ± 2.9%), 10P(34) - = 14.07 $cm^{-1}atm^{-1}$ (error ± 0.48%) and 10P(36) - = 7.39 $cm^{-1}atm^{-1}$ (error ± 0.83%). Compared to the other values reported previously in the literature (Brewer & Bruce, 1978), our measurements indicate a general good agreement.

A high accuracy in measurements was assured by using a frequency stabilized laser (1 MHz laser linewidth), with the absorption coefficients measured only at the top of the laser line, using certified gas mixtures, avoiding saturation and the influence of foreign gases and averaging over several independent measurements at each line to improve the overall accuracy of the results.

The measurements of the absorption coefficients of ethylene and ammonia at CO_2 laser wavelengths are many time a prerequisite for applications involving breath analysis presents usually in the entire field of life sciences.

Chapter 4.
Breath biomarkers analysis at subjects with different pathological issue

4.1. Introduction

Because the exhaled breath air analysis represents an attractive and promising novel approach for noninvasive detection of human biomarkers associated with different diseases, the next chapter is dealing with breath analysis applications developed in the Optics and Lasers in Life Sciences, Environment and Manufacturing Laboratory at National Institute for Laser, Plasma and Radiation Physics, Bucharest, Romania (Popa et al., 2011a, Popa et al., 2011b, Popa et al., 2011c, Popa et al., 2011d, Popa et al., 2013a, Popa et al., 2013b, Popa et al., 2013c, Popa et al., 2014a, Popa C., 2014, Popa et al., 2014c, Popa et al., 2014b)

I characterize by breath test analysis one prevention domain (smoking) and one pathological issue (renal failure), with the assessing of one treatment efficiency (haemodialysis-HD).

People have to know that there is no risk free level of exposure to tobacco smoke and that the ethylene level (reactive gases in the smoke will cause damage and ethylene can be a response from the damage of the human lung tissue) in the Electronic cigarette smoker's case was found to be in smaller concentrations compared to Traditional cigarette smoker's case. Electronic cigarettes may provide a safe alternative to Traditional cigarettes smoking.

Also, people have to know that HD treatment determines simultaneously a large increase of ethylene concentration in the exhaled breath (due to the oxidative stress) and a reduction of the ammonia concentration, correlated to the blood urea nitrogen level. Analysis of ethylene and ammonia traces from breath may provide insight into severity of oxidative stress and metabolic disturbances and give information for determining efficacy and endpoint of HD.

These results demonstrate that LPAS is a sensitive, non-invasive and real time method to accurately analyze breathing ethylene gas concentrations that possess high absorption strengths and a characteristic absorption pattern in the wavelength range of the CO_2 laser.

4.2. Protocol for breath ethylene and ammonia measurements using LPAS method

I have analyzed ethylene and ammonia exhaled from patients receiving HD for treatment of renal disease and ethylene changes at different time intervals in the exhaled breath composition of E-cigarette smokers *vs.* T-cigarette smokers, before and after the consecutive exposure with cigarettes. Breath samples were collected at certain time intervals and the subjects were asked to exhale into sample bags at a normal exhalation flow rate.

After the alveolar air sample is collected, the sample gas is transferred into the PA cell and can be analyzed immediately or later. In either case, it is recommendable to seal the large port with the collection bag port cap furnished with the collection bag. The use of the port cap assures that the sample volume will not be lost due to a leak. Its use also avoids the contamination of the sample by gas diffusion through the one-way valve in the large port, if the sample is stored for a long period of time prior to its analysis.

To ensure the quality of each measurement, an intensive cycle of N_2 washing was performed between samples, in order to have a maximum increase of 10% for the background PA signal. It has to be underlined that the measured photoacoustic signal is due mainly to the absorption of ethylene and ammonia, but some traces of CO_2, H_2O, ethanol, etc., influence the measurements (overall contribution is less than 10%).

To avoid the interference of our molecules of interest with over 700 species of bacteria that live in our mouths, the subjects were instructed to use toothpaste and antiseptic mouthwash before each breath sampling. Also, the response to all absorbing species at a given laser wavelength (photoacoustic signal) decreased considerably when we inserted a KOH trap (with a volume larger than 100 cm^3) proving that amounts of CO_2 and H_2O vapors in the breath can alter significantly the results, thus their removal being compulsory (Bratu et al., 2011).

To analyze the ethylene and ammonia from the bags, we evacuate the extra gas (Popa et al., 2011a, Popa et al., 2011b, Popa et al., 2011c, Popa et al., 2011d, Popa et al., 2013a, Popa et al., 2013b, Popa et al., 2013c) and then we flushed the system with pure nitrogen at atmospheric pressure for few minutes and the exhaled air sample was transferred to the cell at a controlled flow rate of 600 sccm (standard cubic centimeters per minute).

An important parameter in the measurements is the responsivity R (cmV/W) of the PA cell which depends on the pressure of the gas inside the cell. Taking into account the fact that the initial pressure in the sample bags filled by the healthy humans and by the patients/subjects with different disorders differs from one case to other, it is necessary to know the pressure dependence of the PA cell responsivity (Fig. 4.1). The exhaled air sample was transferred to the PA cell at a controlled flow rate of 600 sccm, and the total pressure of the gas in the PA cell was measured, applying then the correction factor for the responsivity according to the calibration curve from Fig. 4.1.

The responsivity of the PA cell was determined by using a calibrated mixture (Linde Gas) of 9.88 ppmV (± 2%) C_2H_4 diluted in nitrogen 6.0 (purity 99.9999%) and of 0.96 ppmV (± 5%) C_2H_4 diluted in nitrogen 5.0 (purity 99.999%) (Dumitras et al., 2011a, Dumitras et al., 2011b). The pressure dependence of the responsivity was measured always at the center of the CO_2 laser line by using a frequency stabilized laser (instability 3×10^{-8}).

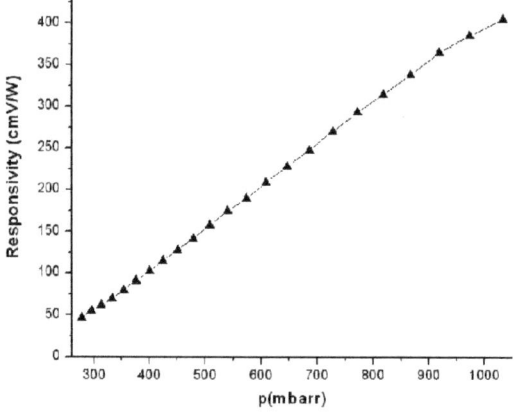

Figure 4.1 The responsivity of the PA cell against the pressure.

The absorption coefficients of ethylene and ammonia (presented in Chapter 3) at different CO_2 laser wavelengths were precisely measured previously (Dumitras et al. 2007, Dumitras et al. 2011, Popa et al. 2011a) and the CO_2 laser was kept tuned at the 10P (14) line (10.53 μm) where ethylene exhibit a strong peak, corresponding to an absorption coefficient of 30.4 cm^{-1}atm^{-1} and at 9R(30) CO_2 laser line (9.22 μm), where the ammonia absorption coefficient has the maximum value of 57 cm^{-1}atm^{-1}.

4.3. Exhaled breath biomarkers analysis from the patients with renal failure treated with HD

4.3.1. Breath ammonia concentration *vs.* blood urea concentration at patients with kidney failure

Ammonia biomarker was measured using the LPAS system (described on Chapter 2) and subjects were recruited from patients receiving dialysis treatment at the renal dialysis clinics at the IHS Fundeni, Bucharest.

Experimental determinations in order to detect traces of ammonia and also to measure the urea level were performed for a healthy volunteer and for 13 patients with kidney failure (Popa et al., 2011b, Popa et al., 2013c).

Analysis of pre-dialysis urea level and post-dialysis urea level were made at MedCenter, Bucharest and the results are presented in Fig. 4.2.

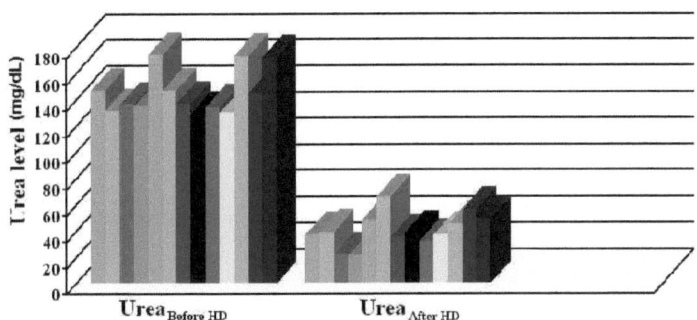

Figure 4.2. Urea data measured for 13 patients with kidney failure.

The exhaled air samples were collected before, and after the dialysis procedure stoped. We have analyzed ammonia exhaled from patients receiving HD for treatment of kidney failure.

Experimental measurements of breath ammonia concentrations for the patients with renal failure and for the healthy subject were performed and the results are presented in Fig. 4.3. The control value for breath ammonia was 0.25 ppm.

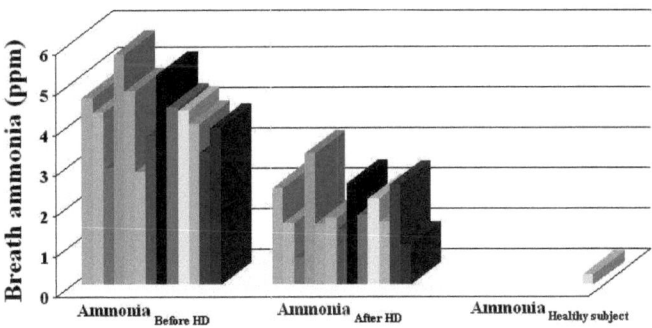

Figure 4.3. Ammonia concentration for 13 patients with kidney failure.

31

In Fig. 4.3, we observed a reduction of ammonia concentration in exhaled breath at patients under HD treatment, which means that ammonia detection in human breath using LPAS system can be used for determining the exact time necessary for the desired state of HD for a patient with kidney failure at every session and, in the same time, could serve as a broad noninvasive screen for incipient kidney disease.

We can see also a remarkable positive correlation between urea data from Fig. 4.2. and the breath ammonia concentration from Fig. 4.3.

The ammonia test is noninvasive, easily repeated, and does not have the discomfort or embarrassment associate with blood and urine tests.

These measurements (Popa et al., 2011b, Popa et al., 2013c, Popa et al., 2013b) were possible because of the high sensitivity of our CO_2 LPAS system, sensitivity that was obtained through successively improvements in optics, laser source and electronics (faster response, low noise equipment).

4.3.2 Breath ethylene biormarker in renal failure of elderly patients

In the recent years there is a large increase in the areas related to the developments in prevention of diseases, especially in explaining the role of oxidative stress. The lipid peroxidation (LP) and oxidative stress contributes to morbidity at HD patients. So, it will be relevant to analyze the impact of oxidative stress and its related species (ethylene) immediately after dialysis treatment in order to prevent trauma in renal failure of elderly patients (Popa et al., 2013c, Popa et al., 2013b).

First I tested the ability of the LPAS system to distinguish the subjects assumed to be healthy. In this way, 10 age matched healthy subjects (Hs) served as controls. The control volunteers (considered healthy) were non-smokers, non alcoholic and non diabetic, without kidney and lung diseases, or other known chronical affections. It should be pointed out that the Hs did not receive HD treatment.

The Hs were advised to use antiseptic mouthwash (the same condition applies for the HD patients) before breath sampling, to minimize the risk of contamination issues and to avoid the oral bacteria.

The average concentration of breath ethylene at Hs was 0.0063 ppm (see Fig. 4.4) with a normal exhalation flow rate.

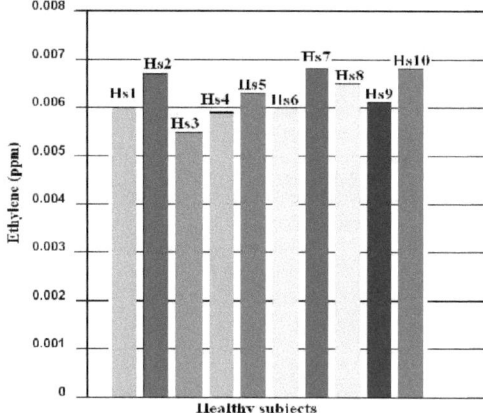

Figure 4.4. Breath ethylene for 10 healthy subjects – Hs.

While much is known about the effect of renal disease and dialysis on the composition of blood, little is known about their impact on the composition of breath. This analysis focuses on the ethylene changes in the exhaled breath composition of elderly HD patients immediately after the treatment of renal disease. Five patients with renal failure (age between 70 and 80 years old) from IHS Fundeni, Bucharest, participated twice in this study, which was designed to explore the effect of urea (from the blood) and the HD treatment on the composition of exhaled breath. Patients had been on HD for periods ranging from 7 to 8 years and were non-smokers, non diabetic and had not consumed alcohol for at least 10 hours prior to their arrival at the clinic. Over a four hour HD procedure (typically per week), breath ethylene was collected from participants before and immediately after the procedure.

For this study the subjects were asked to exhale into sample bags at a normal exhalation flow rate (the same exhalation technique like Hs).

Table 4.1 summarizes the characteristic parameters for ethylene breath test protocol (Popa et al., 2013c, Popa et al., 2013b).

Table 4.1. Essential parameters of ethylene breath test protocol

LPAS system parameters	Specifications
PA cell total volume	≈ 1000 mL
Breath sample flow rate	600 sccm
Working temperature	$23 - 25^0 C$
Breath sample time collection	≈ 10 s
Breath sample time transfer	≈ 2 minutes
Breath sample time analysis	≈ 3 minutes

Figures 4.5 and 4.6 show the experimental results of breath ethylene and urea level for the participants with renal failure before and after HD procedure.

All measurements were made at 10.53 μm CO_2 laser line/10P (14), where the ethylene absorption coefficient has the largest value (30.4 $cm^{-1}atm^{-1}$ at 949.479 cm^{-1}).

Exhaled breath ethylene concentrations in renal failure at elderly patients are considerably changed immediately after the treatment (see figure 4.5), suggesting that subjects are under oxidative stress during HD, and ethylene may be considered a suitable biomarker for LP and oxidative stress in this case.

Because renal failure arises from the inability of the kidneys effectively to work and to clear the blood, an accumulation of urea in the blood is produced; HD treatment helps the patient to remove more urea from the blood. There is a reduction in the urea concentration in the blood of patients as HD proceeds. The results are given in Fig. 4.6.

We have monitored the evolution of the oxidative stress during HD treatment using the exhaled ethylene as a biomarker. We can observe that in elderly patients with renal failure, and particularly in those submitted to HD treatment, the ethylene concentration is increased, proving the existence of reactive oxygen species.

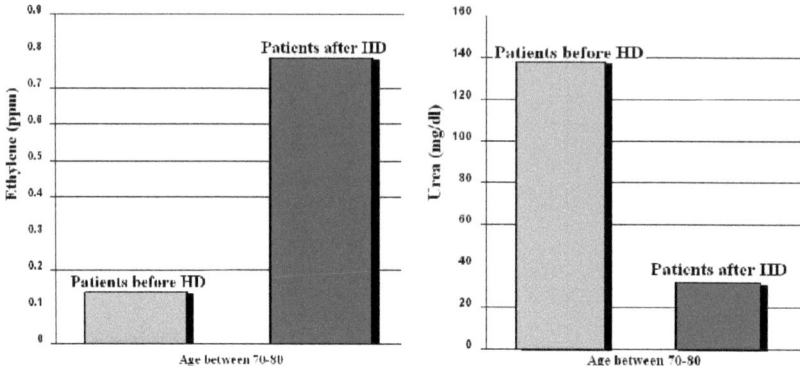

Figure 4.5. The average levels of ethylene - 5 subjects. **Figure 4.6.** The average levels of urea - 5 subjects.

The ethylene concentration in breath is unlikely to be due to a reservoir of ethylene stored in tissue, since ethylene is highly volatile, not significantly metabolized by the body and not soluble in body fat. So, ethylene rapidly diffuses into the bloodstream after generation, being transported to the lungs to be excreted (in the expired breath), from where can be collected (Popa et al., 2013b).

HD may aggravate and intensify the oxidative stress. This seems to be due to multiple factors including an increase in the production of agents from oxidative metabolism, and a decrease in anti-oxidant defenses. Other factors such as the use of low biocompatible membranes and purity of dialysis water, chronic inflammatory state, diabetes, anemia, hypertension, etc., are linked to an increase in LP. Also age, lifestyle and underlying condition affect the overall oxidative stress status of HD patients.

4.3.3 Breath ethylene and ammonia measurements and correlation with urea level from HD patients

HD is a method for extracorporeal removing of the waste products such as creatinine and urea, as well as water from the blood when the kidneys are in kidney failure. HD is one of three renal replacement therapies (the other two being renal transplant and peritoneal dialysis).

HD was accomplished with BAXTER dialysis machines using DICEA (and XENIUM) high performance cellulose diacetate hollow fibre dialyser-gamma series (DICEA 170G) with following characteristics: surface area of 1.7 m^2, ultrafiltration rate 12.5 ml/hr/mmHg, inner diameter of 200 microns and membrane thickness of 15 microns. Experimental measurements in order to detect traces of ethylene and ammonia were performed for a healthy volunteer (C. A. male, 26 years old) and for 6 patients with renal failure.

Participants were recruited from patients receiving HD treatment at the renal dialysis clinics and were dialyzed 3 times per week, with a 4 h dialysis session, instructed to use antiseptic mouthwash before each breath sampling, to avoid oral bacteria.

This time the exhaled air samples were collected before, during (about 1 hour after the start of HD) and immediately after the HD procedure. Experimental measurements of breath ethylene and ammonia concentrations for the patients (P1-P6) with renal failure and for the healthy subject (P0) were performed and the results are presented in Figures 3 and 4, respectively. The control P0 values are 0.006 ppm ethylene and 0.25 ppm ammonia. The details for patients P1 to P6 are introduced in Table 4.2. All measurements were made at 10P(14) CO_2 laser line (10.53 μm), where the ethylene

absorption coefficient has the largest value (30.4 cm^{-1}atm^{-1}), and at 9R(30) CO_2 laser line (9.22 μm), where the ammonia absorption coefficient has the maximum value of 57 cm^{-1}atm^{-1}.
Particular data of patients are summarized in Table 4.2 (Popa et al., 2011b, Popa et al., 2013c).

Table 4.2. The particular data of patients and the experimental measurements of breath ethylene and ammonia concentrations (± 10 % data error)

Patients	Gender	Age	HD since	U_{preHD} (mg/dl)	U_{postHD} (mg/dl)	C_2H_4 (ppm)			NH_3 (ppm)		
						before HD	during HD	after HD	before HD	during HD	after HD
P1	Male	67	2005	147	37	0.03	0.13	0.52	4.63	3.58	2.39
P2	Male	80	2004	131	39	0.23	0.51	0.93	4.28	2.82	1.53
P3	Male	79	2008	136	22	0.17	0.31	0.91	2.89	2.06	0.67
P4	Male	22	2010	135	21	0.14	0.19	0.84	5.71	4.08	3.24
P5	Male	54	2010	174	48	0.18	0.43	0.89	4.79	3.07	1.5
P6	Male	66	2005	147	66	-	-	-	2.8	2.01	1.66

A special mention should be made: NH_3 is a highly adsorbing compound and the results of successive measurements are often altered by the molecules previously adsorbed on the pathway and cell walls. To ensure the quality of each measurement, an intensive cycle of N_2 washing was performed between samples, in order to have a maximum increase of 10% for the background photoacoustic signal. It has to be underlined that the measured photoacoustic signal is due mainly to the absorption of ammonia and ethylene, respectively, but some traces of CO_2, H_2O, ethanol, etc., influence the measurements (overall contribution is less than 10%).

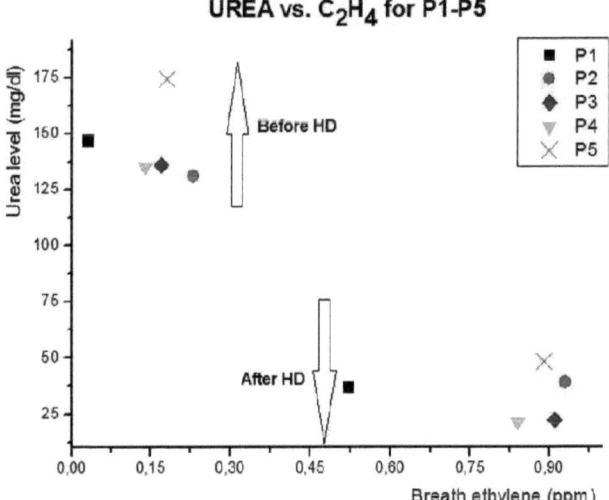

Figure 4.7. Breath ethylene concentration measured for 5 patients with renal failure and correlation with Urea level.

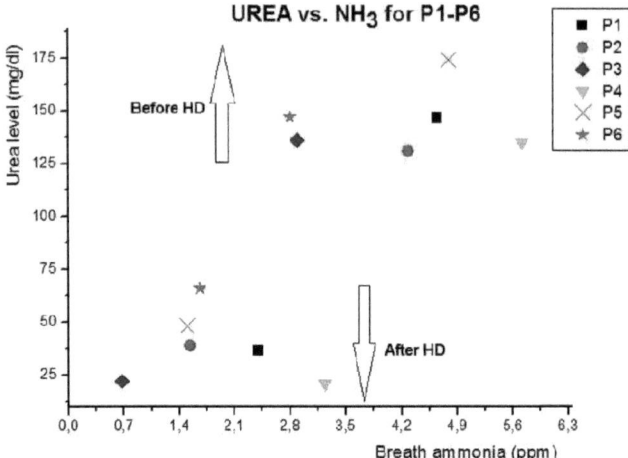

Figure 4.8. Breath ammonia concentration measured for 6 patients with renal failure and correlation with Urea level.

As expected, we see that, (Fig. 4.7), immediately after HD treatment, the ethylene concentration increases, proving the presence LP and showing an imbalance between oxidant and antioxidant systems.

Another biomarker present in patients on HD is ethane which shows absorption at ~ 3.4 m (lead - salt diode laser) and was analysed by Patterson et al. (Patterson et al., 2007) Stevenson et al. (Stevenson et al., 2008) and Handelman et al. (Handelman et al., 2003) in exhaled air during HD treatment. They observed a significant peak in oxidant stress levels and demonstrated endogenous production of ethane by the patient whilst on HD.

HD is associated with increased oxidative stress and all treated patients are exposed to this stress. This observation appears to be due to an increased production of free radicals during HD and immediately after HD and a net reduction of many antioxidants. In order to verify this hypothesis also further studies are required.

In Fig. 4.8, as expected, we observed, a reduction of ammonia concentration in exhaled breath at patients under HD treatment, which means that ammonia detection in human breath using LPAS system can be used for determining the exact time necessary for the desired state of HD for a patient with end stage renal disease at every session and, in the same time, could serve as a broad noninvasive screen for incipient renal disease (Popa et al., 2011b).

The most important result is the correlation found between Urea data (measured by blood analysis) and the individual breath ammonia and ethylene concentrations (measured by photoacoustic technique), shown in Figs. 4.7 and 4.8 for another six patients.

We have found out that the composition of exhaled breath in patients with renal failure contains not only ethylene, but also ammonia and gives valuable information for determining efficacy and endpoint of HD.

Analysis of ethylene and ammonia traces from the human breath may provide insight into severity of oxidative stress and metabolic disturbances and may assure optimal therapy and prevention of pathology at patients on continuous HD.

As future work, we recommend the increasing of antioxidant intake level in HD patients and then compare the oxidative stress production, keeping the present results as reference and the exhaled ammonia/ethylene as specific biomarkers (Popa et al., 2011b).

4.4. Influence of inhaled vapors *versus* inhaled smoke produced by cigarettes

The CO_2 LPAS is suitable for the detection of ethylene in exhaled breath (Popa et al., 2014a, Popa C., 2014, Popa et al., 2014c, Popa et al., 2014b), producing feasible and reproducible results which discriminate active smoking with E-cigarettes (electronic) *vs.* T-cigarettes (traditional).

The E-cigarette closely imitates T-smoking since it tastes, looks and also feels like a traditional one. When "vaping" the E-cigarette inhaling produces both the tactile and craving satisfaction which T-cigarettes seeks and generates a vaporizing process that releases a vapor mist that evaporates into the air within just a few seconds.

Since the introduction of this product to the consumer marketplace, a number of new companies around the world have started producing them in order to take advantage of the overwhelming positive response being generated by the consumer (Popa 2014a, Popa 2014b).

While we can't make the claim that E-cigarettes are healthier, we can point out how T-cigarettes are harmful to our health and can put us at higher risk of a whole host of conditions, including: stroke, heart attack, lung cancer, throat cancer, pneumonia, osteoporosis, Alzheimer's and countless others.

This chapter section reports the LPAS as a sensitive, real time and non-invasive tool to monitor at different time intervals the concentration of ethylene at E-cigarettes smokers and T-cigarettes smokers (Popa 2014a, Popa 2014b).

The data analysis, were conducted for 5 days with ten male's smoker subjects (five of them: smokers only of E-cigarettes and the other five: smokers only of T-cigarettes).

To evaluate the breath ethylene we choose to analyze the effect of the inhalation with E-cigarettes (with 0.5 mg nicotine/drop, 10 mg of nicotine/20 drops) and T-cigarettes (with 0.5 mg /cigarette, 10 mg of nicotine/pack; 0.5 mg x 20 cigarettes) at different time intervals (at 9^{00} a.m. and 10^{00} a.m.) in two sessions.

The subjects were not in the stage of smoking cessation attempt, were non-alcoholic and non-diabetic, without any chronic mental or physical health problem. Also the ten male's smoker subjects were asked, to avoid coffee and alcohol for at least 6 hours prior to their participation in the study and provided three breath samples every day between 8^{30} a.m. and 10^{00} a.m. (at 8^{30} a.m. collection of breath sample before smoking, at 9^{00} a.m. collection of breath sample after the first cigarette and at 10^{00} collection of breath sample after the second cigarette inhalation) over a period of 5 days.

The T-cigarette smoker smoked one cigarette/session/0.5mg nicotine at 9^{00} a.m., with 15-20 puffs/cigarette and 10-15 seconds interpuff interval, during ≈ 10 min smoking session. After that, with a break of about 50 min, the subject start to smoke the second T-cigarette/second in the session (used similarly conditions to the first cigarettes).

In the same time the E-cigarette smoker (similarly to T-cigarette smoker) put one drop with 0.5 mg of nicotine E-liquid in the atomizer and start to inhale with 15-20 puffs/drop, 10-15 second's interpuff interval and ≈ 10 min "vaping" session (see in the fig 4.9). After a break of about 50 min, each smoker repeated the entire session with one E/T cigarette one more times.

The volunteers were asked to smoke the same brand of cigarette to avoid variability in smoke composition (it is known that cigarette from different brands can generate different ethylene levels).

Immediately after the final puff of each cigarette, the smoker exhaled in the sample bag through the mouth. All the volunteers used the same procedure for inhalation of smoke/vapours by cigarettes. All the information's published about the volunteers was the subject to their permission and are provided in Table 4.3 (Popa 2014a, Popa 2014b).

Table 4.3 Subjects information's for T-cigarette smoke and E-cigarette vapours exposure study.

Subject	Gender	Age	Subjects height (m)	Subjects weight (kg)	Smoker since
S1	Male	23	1.81	73.0	2011 (E-cig.)
S2	Male	28	1.83	97.0	2010 (E-cig.)
S3	Male	31	1.62	56.0	2009 (E-cig.)
S4	Male	29	1.79	78.0	2011 (E-cig.)
S5	Male	23	1.68	83.0	2011 (E-cig.)
S6	Male	35	1.78	98.0	2011 (T-cig.)
S7	Male	32	1.78	81.0	2008 (T-cig.)
S8	Male	37	1.93	99.0	2007 (T-cig.)
S9	Male	27	1.65	62.0	2009 (T-cig.)
S10	Male	28	1.84	73.0	2009 (T-cig.)

Figure 4.9. shows the average concentrations of breath ethylene for five subjects, before and after exposure to one E-cigarette/session.

Each breath smoker was investigated for 5 days with 2 exposure session/day, one cigarette/session and about 50 min break between sessions.

The baseline for E-smokers was: 20 ppb (the breath sample was collected before the exposure to E-cigarette: at 8.30 a.m.), and after the first E-cigarette inhalation/Session 1, the mean ethylene level for Subject 1 (S1) was about 47 ppb, for S2: 45 ppb, for S3: 47 ppb, for S4: 53 ppb while for S5 in Session 1: 49 ppb.

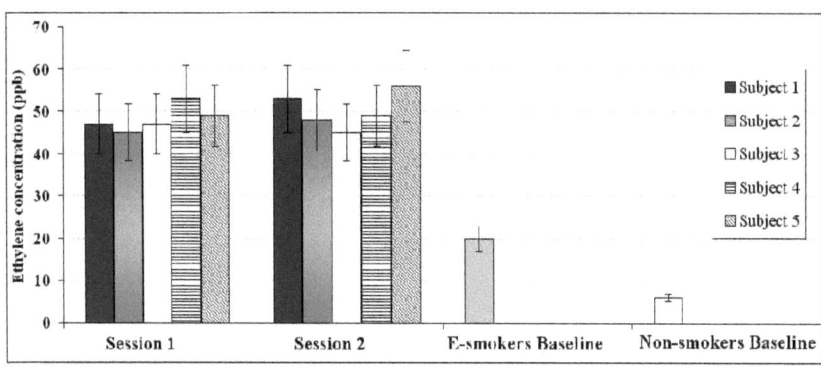

Figure 4.9. Breath ethylene average levels for five E-cigarettes smoker volunteers (Popa 2014b).

For the session 2 the values of ethylene concentrations for the exhaled breath samples were: S1-53 ppb, S2-48 ppb, S3-45 ppb, S4-49 ppb whereas for S5 the value was: 56 ppb.

Figure 4.10 shows the average concentrations of breath ethylene for five T-cigarettes smokers, before and after exposure to T-cigarettes.
The baseline for T-smokers was: 27 ppb (before the exposure to T-cigarette at 8.30 a.m.), and immediately after the first T-cigarette inhalation (Session 1), the mean ethylene level increased for S1 at 145 ppb, following that after the second T-cigarette inhalation (Session2) the mean ethylene concentration to increased more at 187 ppb.

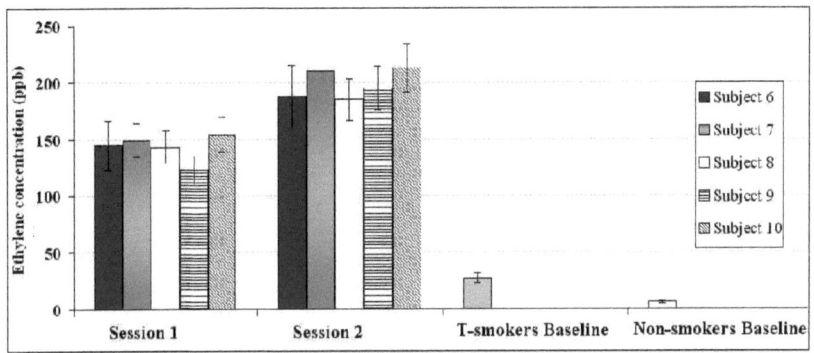

Figure 4.10. Breath ethylene average levels for five T-cigarettes smokers (Popa 2014b).

For the other breath samples in Session 1 and Session 2 exhaled ethylene breath were increased at 149 ppb and 210 ppb for S2, 143 ppb and 185 ppb for S3, 123 ppb and 195 ppb for S4, and for S5 the values are: 154 ppb and 213 ppb.
The results were also compared to the ethylene concentration of a non-smoker subject (6 ppb).
It should be pointed out that the E-cigarettes volunteers did not receive T-cigarettes and the T-cigarettes volunteers did not receive E-cigarettes.
Reactive gases like in the smoke can cause damage and breath ethylene can be a response from the damage of the human lung tissue (the relationship between ethylene and different disease was presented in Chapter 1 section: 1.3).
Based on literature data (Eissenberg, 2010, Gao, 2014, Vansickel, 2010, Bullen, 2010) and compared with our results, we hypothesized that E-cigarettes are safer than T-cigarettes because the ethylene concentration from breath of E-smokers was found to be smaller at different time intervals (9^{00} and 10^{00} a.m.).
In the present study, both the feasibility and the importance of monitoring exhaled ethylene from different subjects have been shown. The ethylene gas, a biomarker of oxidative stress, has been measured using a CO_2 laser based photoacoustic spectrometer.
The results obtained here give the useful information that smoking T-cigarettes, which release in tobacco smoke a complex chemical mixture of combustion compounds (like burned nicotine and tar), causes adverse health outcomes, particularly cancer, cardiovascular and pulmonary diseases, through mechanisms that include DNA damage, inflammation, and oxidative stress.
Oxidative stress from exposure to tobacco smoke has a role in the pathogenic process leading to chronic obstructive pulmonary disease. The evidence on the mechanisms (lipid peroxidation) by which T-smoking causes disease indicates that there is no risk free level of exposure to tobacco smoke.

E-cigarettes (where nicotine is released into vapors) may help reduce smokers' exposure to toxins. Nicotine (while is a highly addictive substance), is not what causes cancer for smokers or for the people around them who breath their second hand smoke.

In summary, the study revealed that E-cigarettes are not so dangereous to cause cancer at smokers, because the ethylene in this case was found to be in smaller concentrations. That is why E-cigarettes may provide an alternative or a substitute to T-cigarettes smoking.

The measurements presented here confirms that the analysis of smokers exhaled breath with CO_2 laser-photoacoustic spectroscopy based instruments is a reliable and non-invasive method, with potential in monitoring the ethylene biomarker from active smoking breath samples.

Conclusions

In the present book, measurements were made to determine if the renal failure and the HD were associated with increased oxidative stress, and to find out if ammonia can be used for selecting the optimum HD duration for a desired state. Ethylene was measured as a biomarker of lipid peroxidation at 5 patients and ammonia was measured at 6 patients, all with renal failure.

Our measurements demonstrated that HD determines simultaneously a large increase of the ethylene concentration in the exhaled breath (owing to the oxidative stress) and a reduction of the ammonia concentration, correlated to the level of blood urea nitrogen.

Another goal of the study book was to determine and verify the evolution of inhalation of vapors and smoke during smoker's subjects.

The levels of ethylene trace gas are much lower after the inhalation of E-cigarette smoker with E-cigarettes at different time intervals compared to inhalation of T-cigarette smoker with T-cigarettes at different time intervals.

The results obtain here could give us the information that inhaling the toxic components of T-cigarettes smoke are deposited in the lung and this has the effect of the activation of endogenous source of free radicals and appearance of oxidative stress together with lipid peroxidation which leads at inflammatory gene activation. The toxic components of T-cigarettes smoke may induce ethylene formation in large quantities.

Breath test is noninvasive, easily repeated, and does not have the discomfort or embarrassment associated with blood and urine tests. Breath is much less complicated mixture than serum or urine and is amenable to complete analysis of all compounds present. No work-up of breath sample is required, in contrast to many analyses performed on serum or urine samples. Breath analysis provides direct information on respiratory function that is not obtainable by other means and can dynamically monitor in real-time the decay of biomarkers in the human organism.

The applications of resonant PA spectroscopy include concentration measurements and trace gas analysis, accurate determinations of thermophysical properties, detections of dynamic processes such as gas mixing or chemical reactions, relaxation processes (determinations of collisional lifetimes of specified quantum states and routes of energy exchange in polyatomic molecules), spectroscopic experiments, studies of aerosols, etc. Trace-gas sensing is a rapidly developing field, in demand for applications such as process and air-quality measurements, atmospheric monitoring, breath diagnostics, biology and agriculture, chemistry, and security and workplace surveillance.

More than 1000 different VOCs including air pollutants originating from the burning of fossil fuels, traffic, or natural sources can be identified and measured with a CO_2 laser based PA instrument. Such studies are prompted by the growing public concern about serious environmental problems such as acid rain, photochemical smog, stratospheric ozone depletion, and global climatic changes.

Breath analysis is a noninvasive medical diagnostic method that distinguishes among more than 1000 compounds in exhaled breath. Many of these compounds, if measured accurately at very low concentration levels, typically in the range of few ppbV, can be used to identify particular medical conditions. Measuring human biomarkers in exhaled breath is expected to revolutionize diagnosis and management of many diseases and may soon lead to rapid, improved, lower-cost diagnosis, which will in turn ensure expanded life spans and an improved quality of life. For example, ammonia levels in the breath can be used to determine the exact time necessary for an optimal degree of dialysis for a patient with end-stage renal disease at every session.

Trace-gas detection techniques based on PA spectroscopy make it possible to discover and control plant physiology mechanisms such as those responsible for germination, blossoming, senescence, ripening, wounding effects, post anaerobic injury, etc. Many agriculturally interesting gases (ethylene, methane, water vapor concentration, carbon dioxide, ammonia, ozone) can be measured *in situ* and in real time with CO_2 and CO laser based photoacoustic spectrometers.

41

In chemistry, PA spectroscopy is useful in the monitoring of chemical processes (reaction rates, equilibrium constants, enthalpies), identification of different compounds (even isomers and radicals), and dimerization of fatty acid vapors.

LPAS system has been demonstrated that it will play an important role in the future of exhaled breath analysis. The key attributes of this technique is sensitivity, selectivity, fast and real time response and ease to use. LPAS system is a sensitive, non-invasive and real time method to accurately analyze breathing ethylene and ammonia gas concentrations that possess high absorption strengths and a characteristic absorption pattern in the wavelength range of the CO_2 laser.

Acknowledgments

I gratefully acknowledge the assistance provided by the Ms. Raluca Popa, Mr. Cornel Popa and Mr. Ionut Achim. In addition, I acknowledge the financial suport of the Sectoral Operational Programme Human Resources Development 2007-2013 of the Ministry of European Funds through the Financial Agreement POSDRU/159/1.5/S/132395.

Role of funding source

This work was funded by a grant from the University Politehnica of Bucharest, Sectoral Operational Programme Human Resources Development 2007-2013 of the Ministry of European Funds through the Financial Agreement POSDRU/159/1.5/S/132395. POSDRU/159/1.5/S/132395 had no further role in study design; in the collection, analysis and interpretation of data; in the writing of the report; and in the decision to submit the book for publication.

References

Angeli, G.Z.; Solyom, A.M.; Miklos, A. & Bicanic, D.D. (1992). Calibration of a Windowless Photoacoustic Cell for Detection of Trace Gases. Anal. Chem., Vol.64, No.2, (January 1992), pp. 155-158, ISSN 0003-2700

Bakouh N, Benjelloun F, Hulin P, Brouillard F, Edelman A, Cherif-Zahar B, Planelles G.(2004), J Biol Chem., Vol. 279, pp.15975–15983.

Beck, S.M. (1985). Cell Coatings to Minimize Sample (NH$_3$ and N$_2$H$_4$) Adsorption for Low-Level Photoacoustic Detection. Appl. Opt., Vol.24, No.12, (June 1985), pp. 1761-1763, ISSN 0003-6935

Bell A.G., (1880), Am. J. Sci., Vol. XX, pp. 305–324.

Bell A.G., (1881), Philos. Mag. J. Sci., Vol. XI, pp. 510–528.

Berkelmans H. W. A., Moeskops B. W. M, Bominaar J., Scheepers P. T. J., and Harren F.J.M. (2003), Toxicology and Applied Pharmacology Vol. 190, pp. 206-213.

Bijnen, F.G.; Reuss, J. & Harren, F.J.M. (1996). Geometrical Optimization of a Longitudinal Resonant Photoacoustic Cell for Sensitive and Fast Trace Gas Detection. Rev. Sci. Instrum., Vol.67, No.8, (August 1996), pp. 2914-2923, ISSN 0034-6748

Bratu M., Popa C., Matei C., Banita S.,. Dutu D. C. A, and Dumitras D. C. (2011), Journal of Optoelectronics and Advanced Materials, Vol. 13, No. 8, pp. 1045-1050.

Brigden, K. and Stringer, R., (2000), Greenpeace Research Laboratories, Department of Biological Sciences, University of Exeter, Exeter, UK., Stop Pollution, p.1.

Bullen C., McRobie H., Thornley S., Glover M., Laugesen M., (2010), Tob Control Vol.19, 98,

Ca W., Duan Y., (2006), Clinical Chemistry Vol. 52, pp. 800-811.

Charles, S.; Batterman, S.; Jia, Atmos. Environ., (2007),Vol. 41, pp.5371-5384.

Coggiola M. J., Oser H., Young S. E., (2004), SRI INTERNATIONAL MENLO PARK CA.

Devasagayam TP, Tilak JC, Boloor KK, Sane KS, Ghaskadbi SS, Lele RD. (2004), J Assoc Physicians India. Vol.52, pp. 794-804.

Dewey, C.F. (1977). Design of Optoacoustic Systems. In Optoacoustic Spectroscopy and Detection, Ch. 3, Y.-H. Pao (Ed.), 47-77, Academic, ISBN 978-0-125-44159-9, New York, NY, USA

Dieter H., (2011), Ammonia, urea production and pH regulation, 2.3.7, p.181.

Dumitras D. C., Bratu A. M. and Popa C., "CO$_2$ Laser Photoacoustic Spectroscopy: Principles", Intech, Croatia (2012); Chapter I in " CO$_2$ Laser-Optimisation and Application", ISBN 979-953-307-712-2, Ed. D. C. Dumitras a.

Dumitras D. C., Bratu A. M. and Popa C., "CO$_2$ Laser Photoacoustic Spectroscopy: Instrumentation and Applications", Intech, Croatia (2012); Chapter II in " CO$_2$ Laser-Optimisation and Application", ISBN 979-953 307-712-2, Ed. D. C. Dumitras b.

Dumitras, D.C.; Dutu, D.C.; Comaniciu, N.; Draganescu, V.; Alexandrescu, R. & Morjan, I. (1981) Frequency Stabilized CO$_2$ Laser Design. Rev. Roum. Phys., Vol.26, No.5, pp. 485-498, ISSN 1221-1451

Dumitras, D.C.; Dutu, D.C.; Draganescu, V. & Comaniciu, N. (1985). Frequency Stabilization of CO$_2$ Lasers. Preprint LOP-55, CIP Press, Bucharest, Romania

Dumitras, D.C.; Dutu, D.C.; Matei, C.; Magureanu, A.M.; Petrus, M. & Popa, C. (2007a). Improvement of a Laser Photoacoustic Instrument for Trace Gas Detection. U. P. B. Sci. Bull., Series A, Vol.69, No.3, pp. 45-56, ISSN 1223-7027

Dumitras, D.C.; Dutu, D.C.; Matei, C.; Magureanu, A.M.; Petrus, M. & Popa, C. (2007b). Laser Photoacoustic Spectroscopy: Principles, Instrumentation, and Characterization. J. Optoelectr. Adv. Mater., Vol.9, No.12, (December 2007), pp. 3655-3701, ISSN 1454-4164

Dutu, D.C.; Draganescu, V.; Comaniciu, N. & Dumitras, D.C. (1985). Plasma Impedance and Optovoltaic Effect in Sealed-Off CO$_2$ Lasers. Rev. Roum. Phys., Vol.30, No.2, pp. 127-130, ISSN 0035-4090

Eissenberg T., Tob Control 19, 87, (2010).

Fung, K. H. & Lin, H.-B. (1986). Trace Gas Detection by Laser Intracavity Photothermal Spectroscopy. Appl. Opt., Vol.25, No.5, (March 1986), pp. 749-752, ISSN 0003-6935

Gao A., Wu Q., Zhang Y., Miao Y., Song C., (2014), Chin. Sci. Bull. Bioanalysis 59 (11):1113-1122. Vansickel R., Cobb C. O., Weaver M. F., Eissenberg T. E., (2010)., Cancer Epidemiology, Biomarkers &Prevention Vol.19, pp. 8.

Gerlach, R. & Amer, N.M. (1980). Brewster Window and Windowless Resonant Spectrophones for Intracavity Operation. Appl. Phys.A, Vol.23, No.3, (November 1980), pp. 319-326, ISSN 0947-8396

Handelman G. J., Rosales L. M., Barbato D., Luscher J., Adhikarla R., Nicolosi R. J., Finkelstein F. O., Ronco C., Kaysen G. A., Hoenich N. A., Levin N. W.,(2003) Free Radical Biology & Medicine **35**, 17.

Harren Frans J.M., Mandon Julien, Cristescu Simona M., (2012) Encyclopedia of Analytical Chemistry, Online, 2006–2012 John Wiley & Sons, Ltd. DOI: 10.1002/9780470027318.a0718.pub2.

Harren, F.J.M.; Bijnen, F.G.C.; Reuss, J.; Voesenek, L.A.C.J. & Blom, C.W.P.M. (1990a). Sensitive Intracavity Photoacoustic Measurements with a CO_2 Waveguide Laser. Appl. Phys. B, Vol.50, No. 2, (February 1990a), pp. 137-144, ISSN 0946-2171

Harren, F.J.M.; Reuss, J.; Woltering, E.J. & Bicanic, D.D. (1990b). Photoacoustic Measurement of Agriculturally Interesting Gases and Detection of C_2H_4 Below the ppb Level. Appl. Spectrosc., Vol.44, No.8, (September 1990), pp. 1360-1368, ISSN 0003-7028

Harvey RA, & Ferrier DR (2011). Lippincott's Illustrated Reviews:Biochemistry Fifth Edition. Wolters Kluwer Lippincott Williams & Wilkins Health.

Hess, P. (1983). Resonant Photoacoustic Spectroscopy, In Topics in Current Chemistry, Vol.111, F.L. Boschke (Ed.), 1-32,

Int. J. Mass Spectrom., Vol. 281, pp. 92-96.

Jae Kwak, George Preti, (2011). Current Pharmaceutical Biotechnology, Vol. 12, pp. 1067-1074.

Jain, (2010). The Handbook of Biomarkers (Humana Press), Chapter 2, pp.23-72, ISBN: 978-1-60761-684-9

Karbach, A. & Hess, P. (1985). High Precision Acoustic Spectroscopy by Laser Excitation of Resonant Modes. J. Chem. Phys., Vol.83, No.3, (August 1985), pp. 1075-1084, ISSN 0021-9606

Kennedy G, Spence VA, McLaren M, Hill A, Underwood C. & Belch JJF (2005), Free radical biology & medicine Vol. 39, pp. 584–9.

Kerr E.L., Atwood J.G.,(1968), Appl. Opt., Vol. 7, pp. 915–921.

Knight JA. (2000), Ann Clin Lab Sci,Vol. 30, pp. 145-58.

Koch, K.P. & Lahmann, W. (1978). Optoacoustic Detection of Sulphur Dioxide Below the Parts per Billion Level. Appl. Phys. Lett., Vol.32, No.5, (March 1978), pp. 289-291, ISSN 0003-6951

Kocielnik R, Sidorova N, Maggi FM, Ouwerkerk M, and Westerink JHDM., (2013). Smart technologies for long-term stress monitoring at work. CBMS, IEEE, pp. 53-58.

Kreuzer L.B., (1971) J. Appl. Phys., Vol. 42, pp. 2934–2943

Kritchman, E.; Shtrikman, S. & Slatkine, M. (1978). Resonant Optoacoustic Cells for Trace Gas Analysis. J. Opt. Soc. Am., Vol.68, No.9, (September 1978), pp. 1257-1271, ISSN 1084-7529

Le Marchand L., Wilkens L. R., Harwood P., Cooney R. V., (1999), Environmental Health Perspectives Vol. 98, pp.199–202 .

Luft K.F., (1943), Z. Tech. Phys., Vol. 5, pp. 97–104

McCurdy M. R., Bakhirkin Y., Wysocki G., Lewicki R., Tittel F. K.,(2007), Journal of Breath Research, Vol. 1, p.12.

Miekisch W., Schubert J. K., Gabriele F.E., Noeldge-Schomburg, (2004), Clinica Chimica Acta Vol. 347, pp. 25-39.

Miklós, A. & Lörincz, A. (1989). Windowless Resonant Acoustic Chamber for Laser-Photoacoustic Applications. Appl. Phys. B, Vol.48, No.3, (April 1989), pp. 213-218, ISSN 0721-7269

Miklós, A.; Hess, P. & Bozoki, Z. (2001). Application of Acoustic Resonators in Photoacoustic Trace Gas Analysis and Metrology. Rev. Sci. Instrum. Vol.72, No.4, (April 2001), pp. 1937-1955, ISSN 0034-6748

Murtz M.,(2005), Optics and Photonics News, Vol.16, pp. 30-35.

Nägele, M. & Sigrist, M.W. (2000). Mobile Laser Spectrometer with Novel Resonant Multipass Photoacoustic Cell for Trace-Gas Sensing. Appl. Phys. B, Vol.70, No.6, (June 2000), pp. 895-901, ISSN 0946-2171

Narasimhan L.R., Goodman W., Kumar N. Patel C., (2001) Proceedings of the National Academy of Sciences,Vol. 98, p.4617.

Nodov, E. (1978). Optimization of Resonant Cell Design for Optoacoustic Gas Spectroscopy (H-Type). Appl. Opt., Vol.17, No.7, (April 1978), pp. 1110-1119, ISSN 0003-6935

O'Hara, M.E.; Clutton-Brock, T.H.; Green, S.; O'Hehir, S.; Mayhew,C.A.(2009)

Olaffson, A.; Hammerich, M.; Bülow, J. & Henningsen, J. (1989). Photoacoustic Detection of NH_3 in Power Plant Emission with a CO_2 Laser. Appl. Phys. B, Vol.49, No.2, (August 1989), pp. 91-97, ISSN 0946-2171

Olafsson, A.; Hammerich, M. & Henningsen, J. (1992). Photoacoustic Spectroscopy of C_2H_4 with a Tunable CO_2 Laser. Appl. Opt., Vol.31, No.15, (May 1992), pp. 2657-2668, ISSN 0003-6935

Patel C.K.N., Burkhardt E.G., Lambert C.A., (1974), Science, Vol. 184, pp. 1173–1176.

Patterson C. S. , McMillan L. C., Stevenson K., Radhakrishnan K., Shiels P. G., Padgett M. J., Skeldon K. D., (2007), Journal Breath Research **1**, 026005, 8pp

Pfund A.H., (1939), Science, Vol.90, pp.326–327.

Phillips, M.; Boehmer, J.P.; Cataneo, R.N.; Cheema, T.; Eisen, H.J.; Fallon, J.T.; Fisher, P.E.; Gass, A.; Greenberg, J.; Kobashigawa, J.; Mancini, D.; Rayburn, B.; Zucker, M.J. (2004). J. Heart Lung Transplant., Vol. 23, pp. 701-708.

Phillips, M.; Cataneo, R.N.; Cummin, A.R.; Gagliardi, A.J.; Gleeson, K.; Greenberg, J.; Maxfield, R.A.; Rom, W.N. (2003). Chest, Vol. 123, pp. 2115-2123.

Poli, D.; Carbognani, P.; Corradi, M.; Goldoni, M.; Acampa, O.; Balbi, B.; Bianchi, L.; Rusca, M.; Mutti, A. Respir. Res., (2005), Vol.71, pp. 1-10.

Popa C, (2014a), accepted to be published in UPB.

Popa C., (2014), Rom. Rep. Phys. Vol. 66, No.4, accepted to be published.

Popa C., (May 2014b) , Journal of Biomedical Optics Vol. 20 No.5.

Popa C., B ni . , Bratu A. M., Pachia M., Matei C., and Dumitra D. C., (2013b), Laser Physics, Vol. 23, No. 12, doi:10.1088/1054-660X/23/12/125701, Article ID 125701

Popa C., Bratu A. M., Dumitra D. C., Pachia M., and B ni , (2014b), Journal of Optoelectronics and Advanced Materials, Vol. 16, No. 1-2, pp. 82-86.

Popa C., Bratu A. M., Matei C., Cernat R., Popescu A., and Dumitras D.C., (2011c), Laser Physics, Vol. 21, No. 7, pp: 1336-1342,

Popa C., Bratu A. M., Ramona Cernat, Dutu Doru C. A., Banita S. and Dumitras Dan C., (2011a), U. P. B. Sci. Bull., Series A, Vol. 73, No. 2, pp.167-174.

Popa C., Dumitra D. C., Pachia M., and B ni S.,(2014c), Laser Physics, Vol. 24, No. 10, doi:10.1088/1054-660X/24/10/105702.

Popa C., Du D. C. A., Cernat , C. Matei R., Bratu A. M., B ni and Dumitra D. C., (2011b), Appl. Phys. B, Vol. 105, No. 3, pp. 669-674

Popa C., Matei C., (2011d), Optoelectronics and Advanced Materials - Rapid Communications, Vol. 5, No.11 ,

Popa C., Pachia M., B ni S., and Dumitra D. C., (2013a), J. Spectroscopy, Article ID 602434, http://dx.doi.org/10.1155/2013/602434,

Popa C., Pachia M., B ni , and Dumitra D. C., (2013c), Rev. Roum. Chim, Vol. 58, No. 9-10, pp. 779-784.

Popa C., Verga N., Pachia M., B ni . , Matei C., and Dumitra D. C., (2014a) , Rom. Rep. Phys, Vol. 66, No. 1, pp. 120-126

Preece W.H, (1881) Proc. R. Soc. Lond., Vol.31, pp. 506–520.

Puiu A, Giubileo G, Addolorato G, Revelli L, Gasbarrini G, Bellantone R, D'Amore A, Lombardi CP, Carrozza C. (2007), Laser Physics, Vol. 17, pp.772.

Pushkarsky, M.B.; Weber, M.E.; Baghdassarian, O., Narasimhan, L.R. & Patel, C,K.N. (2002). Laser-Based Photoacoustic Ammonia Sensors for Industrial Applications. Appl. Phys. B, Vol.75, No.4-5, (April 2002), pp. 391-396, ISSN 0946-2171

Risby, T.H.(2008) J. Breath Res., Vol. 2, pp. 030302.

Rontgen W.C., (1881) Ann. Phys. Chem., Vol. 1, pp.155–159.

Rooth, R.A.; Verhage, A.J.L. & Wouters, L.W. (1990). Photoacoustic Measurement of Ammonia in the Atmosphere: Influence of Water Vapor and Carbon Dioxide. Appl. Opt., Vol.29, No. 25, (September 1990), pp. 3643-3653, ISSN 0003-6935

Ryter S. W., Sethi J. M., (2007), Journal of Breath Research Vol.1, pp. 026004.

Schubert, J.K.; Miekisch, W.; Birken, T.; Geiger, K.; Nöldge- Schomburg, G.F. (2005). Biomarkers, Vol.10, pp.138-152.

Skeldon KD, McMillan LC, Wyse CA, Monk SD, Gibson G, Patterson C, France T, Longbottom C, Padgett MJ., (2006), Respiratory Medicine, Vol. 100 , pp. 300-306.

Springer, ISBN 3-540-16403-0, Berlin, Germany

Stephen P. and James B., (1998), The New England Journal of Medicine, Vol. 338, p.1428.

Stevenson K. S., Radhakrishnan K., Patterson C. S., McMillan L. C., Skeldon K.D., Buist L., Padgett M. J. and Shiels P. G., (2008), Journal Breath Research **2**, 026004, 8pp

Tam, A.C. (1986). Applications of Photoacoustic Sensing Techniques. Rev. Mod. Phys., Vol.58, No.2, (April-June 1986), pp. 381-431, ISSN 0034-6861

Tonelli, M.; Minguzzi, P. & Di Lieto, A. (1983). Intermodulated Optoacoustic Spectroscopy. J. Physique (Colloque C6), Vol.44, No.10, (October 1983), pp. 553-557, ISSN 0449-1947

Tyndall J.,(1881), Proc. R. Soc. Lond., Vol. 31, pp. 307–317.

Viegerov M.L. (1938), Comptes Rendus (Doklady) de l'Acad'emie des Sciences de l'URSS, Vo. XIX, pp. 687–688.

Wang C., Sahay P., (2009), Sensors Vol. 9, pp. 8230-8262.

Weiner ID and Verlander JW., (2011). Am J Physiol Renal Physiol. Vol. 300, F11–F23.

Zharov, V.P. & Letokhov, V.S. (1986). Laser Optoacoustic Spectroscopy, Vol.37, Springer, ISBN 978-3-540-11795-4, Berlin, Germany

Printed by Books on Demand GmbH, Norderstedt / Germany